RECONFIGURING GLOBAL CLIMATE GOVERNANCE IN NORTH AMERICA

Global mechanisms for climate governance have not attained significant international cooperation towards curbing emissions, increasing the likelihood of severe impacts from climate change worldwide. This book provides a fresh look at operational and innovative subnational agreements that have proved to achieve cooperation across regions, inspiring a new style of negotiations capable of effective results in the face of climate change.

María Eugenia Ibarrarán, Universidad Iberoamericana Puebla, Mexico

One of the great questions regarding climate change governance in North America —but also elsewhere—is how to achieve a "scaling up" of multiple and diverse mitigation initiatives, to provide more coordinated and effective policy results. Marcela López-Vallejo locates the seeds of such coordination at the subnational level in the kind of transregional linkages exemplified by RGGI and WCI. What is unique about this analysis is her multi-layered conception of region, one which involves a nesting of governance and policy linkages, social construction and physical connection, as well as economic and energy associations that operate across scales. Through a thorough analysis of these two experiments, López-Vallejo shows us what factors are critical to create the kind of linkages necessary for more synchronized and successful climate change governance—and these lessons are applicable far beyond North America.

Debora L. VanNijnatten, Wilfrid Laurier University, Canada

The International Political Economy of New Regionalisms Series

The International Political Economy of New Regionalisms series presents innovative analyses of a range of novel regional relations and institutions. Going beyond established, formal, interstate economic organizations, this essential series provides informed interdisciplinary and international research and debate about myriad heterogeneous intermediate level interactions.

Reflective of its cosmopolitan and creative orientation, this series is developed by an international editorial team of established and emerging scholars in both the South and North. It reinforces ongoing networks of analysts in both academia and think-tanks as well as international agencies concerned with micro-, meso- and macro-level regionalisms.

Recent titles in the series (continued at the back of the book)

Re-mapping the Americas
Trends in Region-making
Edited by W. Andy Knight, Julián Castro-Rea and Hamid Ghany

Exploring the New South American Regionalism (NSAR)
Edited by Ernesto Vivares

China's Diplomacy in Eastern and Southern Africa
Edited by Seifudein Adem

Regionalism and Regional Security in South Asia
The Role of SAARC
Zahid Shahab Ahmed

Reconfiguring Global Climate Governance in North America

A Transregional Approach

MARCELA LÓPEZ-VALLEJO
Universidad Popular Autónoma del Estado de Puebla, Mexico

Routledge
Taylor & Francis Group
LONDON AND NEW YORK

First published 2014 by Ashgate Publishing

Published 2019 by Routledge
2 Park Square, Milton Park, Abingdon, Oxon OX14 4RN
52 Vanderbilt Avenue, New York, NY 10017

Routledge is an imprint of the Taylor & Francis Group, an informa business

First issued in paperback 2016

British Library Cataloguing in Publication Data
A catalogue record for this book is available from the British Library

The Library of Congress has cataloged the printed edition as follows:
López-Vallejo Olvera, Marcela
Reconfiguring global climate governance in North America : a transregional approach / by Marcela López-Vallejo.
p. cm. – (The international political economy of new regionalisms series)
Includes bibliographical references and index.
ISBN 978-1-4724-1036-8 (hardback)
1. Climatic changes–Government policy–North America.
2. Global warming–Government policy–North America. 3. Environmental policy–North America. 4. Enivornmental protection–International cooperation. 5. Global warming–International cooperation. 6. Climatic changes–International cooperation. 7. Regional Greenhouse Gas Initiative. 8. Western Climate Initiative. I. Title.
GE190.N7L67 2014
363.738'74561097–dc23

2013039239

ISBN 978-1-4724-1036-8 (hbk)
ISBN 978-1-138-27071-8 (pbk)

Contents

List of Tables

List of Figures

List of Abbreviations

ACES	American Clean Energy and Security Act
APEC	Asia-Pacific Economic Cooperation
AZ	Arizona
BC	British Columbia
BECC	Border Environment Cooperation Commission
BRIC	Brazil, Russia, India, China
CA	California
CARB	California Air Resources Board
CC	Compliance Committee
CCCR	Climate Change City Registry
CCS	carbon capture and storage
CDM	Clean Development Mechanism
CEC	Commission for Environmental Cooperation
CFCs	chlorofluorocarbons
CFE	Comisión Federal de Electricidad
CH_4	methane
CITSS	Compliance Instrument Tracking System Service
CLOB	central limit order book
CMP	Conference Meetings of the Parties
CO_2	carbon dioxide
CO_2-BTP	CO2 Budget Trading Program
COP	Conference of Parties
CRE	Comisión Reguladora de Energía
CT	Connecticut
CUSFTA	Canada–US Free Trade Agreement
D	Democrat
DE	Delaware
EPA	Environmental Protection Agency
ERCOT	Electric Reliability Council of Texas
ERU	emission reduction unit
ETS	Emissions Trading Scheme
EU	European Union
FERC	Federal Energy Regulatory Commission
G8	Group of 8
G20	Group of 20
GATT	General Agreement on Tariffs and Trade (later WTO)
GDP	gross domestic product

GEF	Global Environmental Fund
GHGs	greenhouse gases
HCFCs	hydrochlorofluorocarbons
HEAT+	Harmonized Emissions Analysis Tool
HFCs	hydrofluorocarbons
HR	House Ruling
I	Independent
ICLEI	International Council for Local Environmental Initiatives
ICSU	International Council of Scientific Unions
IGCC	integrated gasification combined cycle
IA	Iowa
IL	Illinois
IN	Indiana
IPCC	Intergovernmental Panel on Climate Change
ISO	International Standards Organization
JIM	Joint Implementation Mechanism
JISC	Joint Implementation Supervisory Committee
KP	Kyoto Protocol
KS	Kansas
LG	local government
LULUCF	Land-use Change and Forestry
MA	Massachusetts
MB	Manitoba
MD	Maryland
ME	Maine
MGGRA	Midwestern Greenhouse Gas Accord
MI	Michigan
MN	Minnesota
MT	Montana
N_2O	nitrous oxide
NAAEC	North American Agreement on Environmental Cooperation
NADBank	North American Development Bank
NAEWG	North American Energy Working Group
NAFEC	North American Fund for Environmental Cooperation
NAFTA	North American Free Trade Agreement
NEB	National Energy Board (Canada)
NERC	North American Electric Reliability Corporation
NGO	non-governmental organization
NH	New Hampshire
NH_3	ammonia
NH_{4+4}	ammonium
NIR	National Inventory Report
NJ	New Jersey
NM	New Mexico

NO	nitric oxide
NO_{3-3}	aluminum nitrate
NPCC	Northeast Power Coordinating Council
NY	New York
OECD	Organisation for Economic Co-operation and Development
OH	Ohio
ON	Ontario
OPEC	Organization of Petroleum Exporting Countries
OR	Oregon
PCT	Pacific Carbon Trust
PEMEX	Petróleos Mexicanos
PFCs	perfluorocarbons
PM	Prime Ministerparticulate matter
QB	Québec
RAO	Regional Administrative Organization
REDD	Reduction of Emissions from Deforestation and Degradation
RGGI	Regional Greenhouse Gas Initiative
RI	Rhode Island
RMU	removal unit
RPS	Renewable Portfolio Standard
S	Senate
SBI	Subsidiary Body for Implementation
SBSTA	Subsidiary Body for Scientific and Technological Advice
SD	South Dakota
SEMARNAT	Secretaría de Medio Ambiente y Recursos Naturales
SF_6	sulfur hexafluoride
SO_2	sulfur dioxide
SO_{4-2}	sulfate ion
SPP	Security and Prosperity Partnership (for North America)
TGER	transnational green-economic region
UCLG	United Cities and Local Governments
UK	United Kingdom
UN	United Nations
UNEP	United Nations Environment Program
UNFCCC	United Nations Framework Convention on Climate Change
US	United States of America
USGRP	United States Global Change Research Program
UT	Utah
VT	Vermont
WA	Washington
WCI	Western Climate Initiative
WECC	Western Electricity Coordinating Council
WMO	World Meteorological Organization
WTO	World Trade Organization

List of Orders of Magnitude

b	barrel
bbl/d	billion barrels per day
Btu	British thermal units
CAD	Canadian dollars
CO_2e	carbon dioxide equivalent
D	day
g	gram
H	hour
J	joule
kW	kilowatts
kWh	kilowatts per hour
m^2	square meter
m^3	cubic meter
mbl/d	million barrels per day
MW	megawatts
PM2.5	particulate matter up to 2.5 micrometers
PM10	particulate matter up to 10 micrometers
PPBV	parts per billion by volume
PPM	parts per million
PPMV	parts per million by volume
T	tonne (metric) = 1000kg
USD	US dollars
W	watt
Yr	year

Foreword

We live in a fascinating era in which technological change has made possible the emergence of non-conventional and renewable energy assets, and climate change awareness has spurred new mechanisms for encouraging transnational cooperation in order to mitigate greenhouse gases (GHGs) and other pollutants to ensure a better governance of environment and world resources. In spite of this, the implementation of global mechanisms for abating GHGs, such as the Kyoto Protocol, has failed and alternative governance mechanisms seem to emerge at the local sphere. This is one of the great findings that the author Marcela López-Vallejo communicates to the reader. While explaining the flaws and failure of the multilateral mechanism devised at the end of the 1990s in Kyoto, the author illustrates in parallel how "clean air" began to be provided by local government and authorities in North America. Her book accounts for a thorough analysis of two major "transnational green-economic regions," as she calls them, which loomed on the North American horizon at the end of the past decade. By introducing a carbon market approach and a cap-and-trade mechanism for abating pollutants at local government levels (state and provincial), the Regional Greenhouse Gas Initiative (RGGI) and the Western Climate Initiative (WCI), spurred transnational cooperation among key regions of the United States and Canada in order to fulfill what the Kyoto Protocol failed to accomplish, the supply of a "public good"—clean air—though with a different starting point from national or global governance. Marcela López-Vallejo argues that this prototype of transnational green collaborative effort could become the building blocks for encouraging green cooperation with a multi-layer approach and a transregional scope.

While Marcela López-Vallejo's study focuses on new ways of environmental governance, her book is also a contribution to the debate on the present and future of North American integration. Since the inception of the North American Free Trade Agreement (NAFTA), in 1994, the discussion about the mechanisms driving transnational collaboration and integration among NAFTA partners (Canada, Mexico and the United States) became dominated by economic analysis and, to a lesser extent, organizational approaches. At the turn of the second decade of this century, these types of studies have diagnosed a crisis and stagnation in the integration process. Needless to say, those approaches have privileged the inter-state agenda of conflict and collaboration among the three regional partners. By decentering the focus of analysis from the national state to the local level of authority and the geography of social organization, Marcela López-Vallejo's book contributes greatly to a different understanding of regional integration in North America, offering, at the same time, a more optimistic outlook for the region.

Written in clear and fluid prose, with an educational and explanatory intention, this book is necessary for anyone interested in the governance of climate change, as well as for those studying and working on alternative North American policy architectures conveying cooperation and outcomes-oriented solutions. Marcela López-Vallejo demonstrates solid scholarship in the topics she explores, while opening a promising research agenda at theoretical and methodological levels.

ISIDRO MORALES
Director
EGAP, Gobierno y Política Pública
Tecnológico de Monterrey, State of Mexico Campus
Mexico

Acknowledgements

I wish to acknowledge different sets of people who gave advice throughout the process of researching and writing this book. First are those experts who guided the research, particularly professors Isidro Morales Moreno, María Eugenia Ibarrarán Viniegra and Marianne H. Marchand. Other well-known academics deserve a special mention due to their constant academic support: José Luis García Aguilar, Marco Almazán St. Hill, David Mena Alemán, Rafael Velázquez Flores, Jorge Schiavon, Roberto Domínguez, Teresa Gutiérrez-Haces, Ángel de la Vega, Isabel Studer, Debora van Nijnaten, Claude Denis, Delia Montero and Pierre Sved.

The second group of people to acknowledge comprises those who have given institutional support. Firstly, acknowledgements and gratitude go to my host institution, Universidad Popular Autónoma del Estado de Puebla, to the Department of Social Sciences, and especially to Nora Arroyo Carrasco, Coordinator of the School of International Relations, for the support and friendship she has always provided. Secondly, I wish to recognize the support of other academic institutions, such as Universidad de las Américas, Puebla, Benemérita Universidad Autónoma del Estado de Puebla, Université de Montréal, University of Ottawa, and McGill University.

This group also includes networks that have been very important for the academic development of the research, such as Asociación Mexicana de Estudios Internacionales (AMEI), Regionalismo, Desarrollo y Fronteras (REDESFRO), Programa de Investigación Nacional sobre Cambio Climático-UNAM and the Instituto Global para la Sostenibilidad. Within this grouping, I would like to dedicate a special note to the Oxford Institute for Energy Studies, which hosted me as Summer Fellow in 2013 and supported me in every sense through materials, seminars, and, the most important asset, exchange of ideas with the experts there.

The third group is a more personal one. I wish to acknowledge the support and friendship of Mónica Trujillo López for all the helpful advice she has always given me. A special mention goes to Salvador Manilla Lerma for his personal support and involvement in this project. His critical views coming from Science were fundamental in developing the basic premise of the book. Finally, I thank for their constant support my family in Mexico and in the UK, and all my friends and colleagues who gave me their time and advice during the process. I dedicate this book to all of you.

MARCELA LÓPEZ-VALLEJO
Puebla, México—Oxford, United Kingdom

Introduction

> We are the offspring of history, and must establish our own paths in this most diverse and interesting of conceivable universes—one indifferent to our suffering, and therefore offering us maximum freedom to thrive, or to fail, in our own chosen way.
>
> S.J. Gould

Global climate governance, represented by the United Nations Framework Convention on Climate Change (UNFCCC) and the Kyoto Protocol (KP), presents general problems that have led to unsuccessful performance, while opening the door to new transregional governance schemes, especially in North America.

The first challenge concerns a theoretical ambiguity of the concept of governance in relation to authority, legitimacy, and responsibility. Secondly, at a more specific level, there are issues regarding institutional design that have prevented countries from accomplishing the goal of reducing greenhouse gases (GHGs) proposed by the KP.

In order to address the theoretical problems, it is necessary to start analyzing the meaning of global climate governance. Governance has been defined in different times and ways. These definitions have taken various approaches—for example, as social networking outside the state,[1] as the institutionalization of international agenda topics by the United Nations (UN) and international businesses,[2] as a tool for measuring policy in a results-based management approach,[3] as a synonym for policy-making processes establishing that regulation provides the essential link between government and governance,[4] or as policy networking, where there is an intrinsic relation between structures and processes of emerging actors in diverse contexts.[5]

The concept has also been linked to a political neoliberal program expressed in corporate governance and the minimal state.[6] In this sense, governance has been oriented to politics and policy and based on the relations between states and markets where state institutions should guarantee accountability and effectiveness of policies for the market to operate.[7] This approach to governance has been

1 Kooiman, 2000.

2 Mingst, 2004.

3 Durant et al., 2004.

4 Jordan et al., 2005.

5 Marsh and Smith, 2000. See also Dingwerth and Pattberg, 2006; Doornbos, 2004; Weiss, 2000.

6 Viotti and Kauppi, 2010; Rhodes, 1996.

7 Dingwerth and Pattberg 2006; Doornbos, 2004.

contested strongly because it relates to "good" or "bad" governance practices, a situation that led some authors to understand this moral qualification as a mechanism of domination legitimized globally, and usually labeled with the Foucauldian term "governmentality."[8] In the context of contemporary neoliberal institutions, governmentality means that moral judgments are imposed by world leaders to homogenize political practices that are supported by international governance.[9]

In this same sense, governance has been contested for its theoretical universalism.[10] Some scholars mention that, if the concept and the political program were truly universal, they would not lack mechanisms of social participation and implementation of the agreements that global institutions design.[11]

The flexibility that prevents having *one* concept of governance may provide accommodative interpretations. However, this situation leaves room for creative interpretation, portraying numerous benefits. One of the most important is that the concept can be constructed with radiuses of interaction and scope, which change its essential meaning.[12] The aim of this book is to build a radial concept piece by piece, taking governance as a starting point, and then presenting new schemes of governance interaction among actors that go beyond the State or the International Institutions.

The first step is to introduce the environmental dimension into the concept. Historically, the first wave of "environmental governance" was developed in the decade from 1970, when this concept offered solutions to problems such as acid rain. These first concerns became policies that were, in general, regulatory, reactive, focused on single pollutants and on single means to address it (one solution for one problem), and technology driven.[13] At the time, responses to this limited approach emerged; and during the following decades, a new set of definitions and policies appeared, ranging from the need to marketize the environment, to the development of international rules and law, and, finally, to the need to reform the previous command-and-control regulatory frameworks and to democratize them under a communitarian basis. Subsequently, a third wave for understanding environmental governance emerged. This focused on the need for accountability and for results in environmental terms, rather than accountability for compliance with rules and regulations.[14] Simultaneously, the environment was no longer understood as monolithic. A diversity of topics appeared within the environmental area, such

8 This concept was coined by Michel Foucault in 1975. Contemporary scholars tend to use it in a globalization context. See Foucault, 1975.

9 Zanotti, 2005.

10 Krahmann, 2003; Litfin, 2000; Mol, 2002.

11 Jagers and Stripple, 2003.

12 A radial concept has different parts with potential for separate definitions, yet together forming an essential concept not understood without its parts. For further discussion, see Collier and Mahon Jr., 1993.

13 Durant et al., 2004:1.

14 Ibid.:2.

as the depletion of biodiversity, the pollution of water, the lack of appropriate management of waste, and climate change as a product of global warming.

In this context, climate governance appeared as a term of its own during the creation of the UNFCCC in 1992. The launching of the KP during the third meeting of the Convention in 1997 supported the link between climate change and a global response in the hands of the UN institutions. This link was inevitable when thinking of the ultimate goal of climate governance: to provide clean air globally. Clean air started to be conceived as a global common or global public good.[15] This common good needed to be provided by someone other than the markets to avoid market failures and unequal distribution.[16] Although clean air is global, impacts of not having it and adaptation capacities are very different and specific according to the regional particularities and circumstances of the ones polluting it.[17] In other words, in spite of the "globalism" of air, responses needed to have territorial and administrative bases. These responses have been organized according to a state-based approach under national policies as proposed by members of the UNFCCC.

Nonetheless, two decades after the first definitions of climate governance appeared and structures started to work under the auspice of the UN, no clear results have been achieved. For example, some countries in Europe have accomplished their commitments under the KP to date; this is the case of the original 15 members of the European Union (EU). However, they have succeeded only for carbon dioxide (CO_2), whereas the KP deals with five more GHGs. In addition, some of the countries obliged under the KP to limit their GHG emissions have registered the most important reductions immediately after their incorporation into the capitalist markets.[18] Following that period, the amount of GHG reduction has been almost null.

In the region of North America, results have also been unfavorable, and Canada was the only country obliged under the KP to quit this instrument in 2012 when proving incapable of achieving its GHG reduction target of 6 percent. The country increased its GHG emissions by 18 percent from 1990 to 2011.[19] As mentioned in the 2012 Canadian GHG report, steady increases in annual emissions characterized the first 15 years, followed by fluctuating emission levels between 2005 and 2008, and a steep drop in 2009 with emissions somewhat stabilizing in 2010.[20] However, the decrease of the last years is more the result of the economic crises than of climate governance strategies. In the case of the United States (US), in 2001 the Congress decided not to ratify the KP, a decision implying that the country with most GHG emissions worldwide at that time had no commitment within the global climate governance institutions. The US has also increased its emissions,

15 Ostrom, 1990. See also Kaul et al., 2003.

16 Ibarrarán and Boyd, 2007. See also Kaul et al., 2003.

17 Cohen et al., 2004. See also Kaul et al., 2003.

18 This is the case of the ex-Soviet countries of Eastern Europe.

19 Environment Canada, 2012; see also Juillet and Paterson 2009.

20 Environment Canada, 2012; see also Juillet and Patterson, 2009.

by 8.4 percent from 1990 to 2011.[21] After the economic crisis of 2008–2009, GHGs started going up again, by 3 percent to 2012 due to an increase in energy consumption across all sectors as a consequence of the mild economic recovery of its economy.[22] Mexico had ratified the KP in 1998, but as a non-developed country, it was integrated into the "Non-Annex" list, having the only goal of delivering annual reports of the state of GHGs to the UN. Mexico has delivered its reports promptly, but it has also increased its GHG emissions, by 33.5 percent to 2010.[23] Interestingly enough, in North America the country outside the KP scheme and the most developed economic nation worldwide has been doing better in managing GHGs than those under the climate governance of the UNFCCC.

In North America, all three countries are far from accomplishing their original goals (within or outside the KP). To be successful, they would have to cut the use of fossil fuels, which is the leading emitter sector in the region. Clearly, Mexico has been burning a lot more fuel and with a faster speed to develop its economy in the last decade, which is the case of almost all emerging economies.

This scenario evidences that global climate governance institutions have not been able to create incentives for countries to deliver results, as regional or worldwide GHGs have increased. In this sense, there has been a debate whether to measure GHGs on an absolute basis or in another manner. Some countries accuse others of free-riding and not doing enough. For example, China and India argue that, if we were to measure GHG reduction by taking into account the intensity factor (unit of CO_2 per unit of gross domestic product, GDP) or per capita basis, they would not be at the top of the ranking. In addition, according to this way of measurement, no single country would have accomplished its reduction goals.[24]

Why has no country under the KP (worldwide and in North America) accomplished reduction goals? Does this mean that, after lots of money and time invested since 1988, global climate governance has failed and we are worse than when we started? Where does authority need to be located to internalize global climate governance norms successfully and transform them into policies? What are the alternative proposals for so doing? In other words, who can provide public global environmental goods as clean air?

The general failure of global climate governance offers the opportunity to think about alternatives for providing public goods globally and situating localism not as the antithesis of globalism, but as one layer of global processes and relations. Within this framework, new institutional arrangements are being created among different actors with different goals based on different types of power relations, also with the participation of different levels of authority.[25] According to Sverker

21 Environmental Protection Agency, 2012.

22 Environmental Protection Agency, 2012.

23 Instituto Nacional de Ecología, 2012.

24 For acknowledging specific data for each country, see Intergovernmental Panel on Climate Change, 2010.

25 Duchacek, 1988. Another interesting approach is developed by Hocking, 1999.

Jagers and Johannes Stripple, global climate governance should be the object not only of national policies, but also of other authorities and actors.[26] This argument leads to the concept of global governance understood as a policy process in which these new linkages develop in a context where global climate rules need to be internalized partially or completely by different actors in different ways. This "double-level" strategy has been used by actors of global governance especially for climate change.[27] In this sense, governance can be defined by "the fragmentation of authority among actors who may be located in or who represent and regulate different geographical units across national and regional borders."[28]

In this context, the goal of this book is to respond to the previous questions by analyzing alternatives to global climate governance and challenging the premise that, for global problems, the world needs global solutions. The main argument of this book deals with a theoretical concern about the non-obligatory characteristic of the global governance framework. This book proposes that, for accomplishing reductions in GHG emissions, it is necessary to find some other actor with authority for internalizing global problems as policies. In this sense, relocation of authority is essential for providing global public goods such as clean air. To demonstrate this premise, this book analyzes the new institutional arrangements to reterritorialize the provision of global public goods on a transregional basis through emerging carbon markets in North America.

This premise intends to go beyond the link between global public goods and global climate governance in different senses, by presenting a regional treatment for global public goods. This assertion leads to redefining the concept of region as socially constructed in a transnational way, where institutional design needs to be addressed in a dynamic manner for allowing relocation of authority. To perform authority, it is necessary to formulate some sort of leadership role regarding the actors involved. This book proposes that local governments (LGs) can be the main actors of this regionalization process.

The book is divided into seven chapters. The first chapter captures and details previous discussion of the concept of global climate governance and its limits. Through revising the diverse literature, the chapter approaches the concept of governance as a process for designing and implementing policy that coordinates fragmented interests among different actors and levels of authority. Additionally, it presents a discussion of global climate governance and explores the way in which environmental topics were included in the international agenda during the second part of the twentieth century.

Chapter 2 discusses the argument that the global climate governance that is necessary has to deal with the provision of global public goods, such as clean air. Global public goods present several challenges regarding their provision. This chapter argues that, although clean air is a global public good, it needs to be

26 Jagers and Stripple, 2003.

27 Agrawala and Andresen, 2001.

28 Krahmann, 2003:333.

provided at a different level, other than the global or national. The second section of the chapter examines the science of climate change through the lens of Social Sciences. Scientific facts and debates concerning the anthropogenic influence in changing climate patterns have been the basis for granting legitimacy to climate governance institutions to pledge for the provision of global public goods.

The third chapter deals with the failure of the KP as an expression of mainstream global climate governance, asserting that this can be explained in three possible ways. First, the KP is a mechanism with an inadequate design, because it was based on a national approach, whereas the implementation costs were localized regionally and by sector. Second, the KP has no obligatory methods for implementation; it is based on voluntary instruments, targets, and commitments. The third explanation deals with the emergence of free riders. This argument was developed by US Congress to account for its decision not to ratify the KP. Congress explained that potential pollutant countries, such as China and India, which were growing fast and leaving a huge GHG footprint,[29] were not included in the list of countries with actual reduction commitments ("Annex I"), so that the US (and the other Annex I countries) would have to absorb the greatest economic costs.

Chapter 4 explores other levels of authority that could provide global public goods for addressing climate change and its governance. The first part starts by analyzing the regional multinational efforts in a North American context. Regional institutions such as North American agreements on commerce (NAFTA), on environment (NAAEC), and on energy (NAEWG)—not part of the climate global governance framework—had incorporated the topic into their agendas. In North America, as in almost the rest of the world, the sector responsible for the majority of emissions is energy. For this reason, a special section on the status of energy connections in the region is included. Subsequently, the trajectories of national climate and energy policies in the three countries of the region are presented. Needless to say, regional and national policies have not been successful in developing strategies for reducing GHGs either.

Nonetheless, a different regionalization process is taking place in North America at the time of writing. Chapter 5 analyzes how local governments[30]

---

29 China contributes 15 percent of GHGs and India 6 percent, almost the same as Russia (Annex I country). Both countries pollute more than many of the Annex I countries, such as Japan, Canada, and the United Kingdom. See database of the World Climate Report, 2011.

30 The term *local governments* includes all kinds of political governments in different jurisdictions from a nation state, irrespective of the type of political regime to which they belong; it does not matter if they are part of a federated or of a centralized regime. LGs can form part of a state or be divided into two or more countries, as per the Basque Country; LGs can also perform on a micro-scale, such as within municipalities and cities. In this book, LGs refers to the provincial level (encapsulating municipalities and cities) and the membership of a nation state. Iñaki Aguirre offers an excellent semiotic discussion of phenomena as the international action of these kinds of governments and similar concepts of their interaction with other global actors, such as *paradiplomacy*, *postdiplomacy*, or

(provincial or state) need to be considered fundamental actors for redefining global climate change governance and its strategies for reducing GHGs at both an individual and a transregional level. This process implies the creation of regional building blocks based on LGs' integration of policies and leadership. LGs have shown capacities to adapt their governing machinery to build relations with different actors at different levels for responding to global climate pressures that demand implementation of solutions. However, the institutional design for the reconfiguration of climate governance based on a local/regional approach would have to build strong relations between institutions, society, and business. The main goal of this form of transregional governance is the co-production of collective global goods with their citizens.[31] In addition, this type of governance allows creating interrelations with other initiatives of this type and with national and global institutions to take advantage of the opportunities these actors can provide.

Brian Hocking refers to the qualities of local governments[32] that enable their consideration as actors in world politics, and suggests that it is necessary to go beyond the perspective of exclusivity, discontinuity, and separatism in the relation this unit of analysis has with other actors. He explains that where

> [w]e are confronted by an ever widening "frontier" where domestic and foreign policy meshes into a seamless web and in which differing "spheres of authority" jostle one another for advantage, then non-central governments [LGs] can, by virtue of their actorness, be seen as patrolling the edges of this territorially imagined and yet partially deterritorialized space.[33]

James Rosenau had led the discussion of the need to reterritorialize responses for implementing policy solutions for climate change. This process can mean the redefinition of sovereignty regarding climate governance. In this sense, sovereignty is a perforated concept involving, besides national governments, a series of groups like transnational companies, political parties, non-governmental organizations, and territorial communities that initiate or receive trans-sovereign signals, subject to external responses.[34] Nonetheless, LGs' territories appear to be one of the last frontiers with real capacities for actioning policy responses to global pressures, such as climate change.

The last section of Chapter 5 includes a concrete theoretical proposal for reterritorialization and coordination of policies for climate change into new forms of articulation of interests in transnational green-economic regions (TGERs). This

---

constituent diplomacy, as well as definitions on the dichotomies of national–subnational and global–local: see Aguirre, 1999.

31 Hart and Pakrash, 1999.

32 This author uses the concept of *non-central governments* to refer to what here is called local governments.

33 Hocking, 1999:17.

34 Duchacek, 1988:4.

kind of region is conceived as a social construct,[35] with specific objectives that include integration and coherence within economic activities and that sometimes are transnational.[36] This coherence also involves designing integrated sector and climate policies and strategies,[37] and agreeing that the territory is suitable for sustainable development.[38]

TGERs in North America tend to create local carbon markets to accomplish their climate-change goals. However, this does not mean that governance of TGERs leaves aside national or global arrangements. These transregional mechanisms integrate such arrangements into governance instruments, acknowledging that sovereignty is fragmented into sector policies and specific goals and that cooperation with different levels of authority and actors is needed to ensure a substantial reduction in GHG emissions. In other words, a TGER reinvents the borders and scope of the failing global climate governance, where the cornerstone is the relocation of authority and leadership in the hands of LGs.

Chapter 6 provides two case studies of TGERs in North America, the Regional Greenhouse Gas Initiative (RGGI) and the Western Climate Initiative (WCI). RGGI is the result of a merge between the New England Governors and the Eastern Canadian Premiers. To date, it is integrated by ten local governments from the US. WCI was originally integrated by several US states and Canadian provinces, but since 2012, only California and Quebec have continued to work together.[39] In both cases, there are several LGs with the status of observers coming from Canada and Mexico. RGGI and WCI share common approaches to GHG reduction through energy efficiency and emissions trading schemes (carbon markets). In a more specific approach, they have established specific GHG reduction goals, have created institutions, and have designed mechanisms for linking with other initiatives. This chapter provides a detailed explanation of the way their climate instruments work in three aspects: (1) how their cap-and-trade systems function, (2) what instruments they have to implement and achieve reduction goals, and (3) how they create the material conditions to operate the carbon market. In all cases, the electricity sector serves as the main example to evidence integration of policies, institutions, and infrastructure within each of the TGERs.

---

35 Boas et al., 2005.

36 Clarkson, 2008; Morales, 2008; Ramos, 2002.

37 Rowlands, 2008; Bernstein et al., 2008.

38 Toledo and Moguel, 1992.

39 Until 2011, there was another TGER in North America, the Midwestern Greenhouse Gas Reduction Accord (MGGRA). This transregional initiative was built on the basis of the Midwestern Governors Association. MGGRA was established in 2007, and it was integrated by six US LGs (IL, IA, KS, MN, MI, WI) and one Canadian province (MB). By 2010, others joined as observers (IN, OH, SD, and the province of ON). However, MGGRA stopped activities during the fall of 2011. Reasons were diverse. First, all the members withdrew the initiative as a result of their slow recovery from the economic crisis. Second, US mid-term elections of 2010 and Republicans made big gains in the Midwest, dismantling the initiative.

The final chapter of the book provides a comparison of strategies of both TGERs. This comparison also explores what kind of integration they are experiencing and the levels of linkage among them or with other governance actors. This chapter is thus divided into analysis of convergence, harmonization, and linkage criteria for both cases. It seeks to demonstrate how efficient has been the relocation of authority under LGs to address climate governance. In addition, it presents the challenges and opportunities of a transregional approach to climate change in North America.

Limits to this Research

This research is limited by some important factors. RGGI has operated the carbon market for more than a year (holding 19 auctions to date). However, WCI had major problems in beginning operations on time. To date, only one of the members, California, has started the auction process. This situation makes it difficult to gauge a final balance of its achievements overall. Another very important limit is the focus. In the revised literature, governance appears as a policy process in which several actors interact. In the case of this book, the main focus is placed upon only one of the actors in this interaction, LGs. The role of companies and the influence of civil society are included only contextually. Business is considered the principal object of technological change and implementation of GHG reduction. It is important to acknowledge that there are several business mechanisms that push for innovation. An example of this is the International Standards Organization (ISO), which had developed several environmental standards, such as the ISO14000 for climate change or the ISO26000 for social responsibility and environment. Nonetheless, these mechanisms are voluntary and depend on the market and economic trends.

The case of NGOs or other forms of social organization is similar. They influence decision making through democratic participation either in favor of or against policies regarding climate change. Indeed, society is included when addressing consumption patterns regarding climate change-related sectors, as energy. Organized civil society is also taken into account in this book when analyzing how it pushes forward national GHG commitments and policies or global agreements. Civil society appears again when exploring the legitimacy of climate policies, either in favor of or against it. For example, some civil society groups have lobbied to block any local climate policy under the argument that local authorities tend to lower environmental standards. On the other hand, some of these civil society groups have allied with companies to at least force the inclusion of the topic into their day-to-day operations. Both business and civil society agents have their own agendas, which sometimes do not coincide with the provision of public goods.

A final limitation to this book is the scope of a regional approach in North America. The inclusion of Mexican states in transregional carbon markets is still pending. Despite their status as observers, the situation of Mexican states is a difficult one taking into account the constitutional limits to energy sources. Even though Mexican states could participate in a transregional market, they could by no means link their climate strategies to energy policies without the approval of the federal government. However, some of these states have already designed their GHG inventories and other instruments under the TGERs methodologies. Once RGGI and WCI have matured, the results may well attract other LGs and show how efficient this approach may be for North America.

Chapter 1
Constructing the Concept of Global Climate Governance

Introduction

Historically, governance has been defined in different ways, conceived as social actors creating networks, as institutionalization of the international agenda, or erroneously as a synonym of government. However, there are some intermediate approaches that need to be explored to better comprehend the concept. The redefinition of authority, policy processes, norms and institutions is useful to narrow down theories that apparently are too wide. One of the virtues of the concept of governance is its flexibility. The fact that governance has a wide scope of actors, problems and interpretation allows scholars to build on former definitions to construct new ones. Yet this ambiguity can also lead to conceptual flaws where parts of the problem could be neglected.

Therefore the goal of this chapter is to discuss the concept of governance and find elements to develop a useful intellectual tool to work with throughout the book. This chapter intends to narrow down the concept and find its link with climate change. In this sense, climate governance is used as a radial concept composed of different parts that form an essential whole.[1] It is assumed here that governance can be understood as a conceptual tool for measuring policy using a results-based approach.[2] As a consequence of the high costs of command-and-control environmental policies, a new generation of approaches has been emerging. One of these is the results-based approach. For climate governance, this seems appropriate when addressing many diffuse and heterogeneous sources of GHGs (greenhouse gases). The approach includes other actors, chances to establish comprehensive and complementary policies, and to situate pollution geographically.[3]

The results can appear as policy results or as specific issue-area results (as in the case of environmental governance). This is particularly important for the provision of public goods, which is one of the main goals of governance schemes. To obtain this double-track results approach, the construction of governance rules would need to be different from what climate governance has performed to date. Other actors, besides states and international institutions, need to be involved in

1 For a deeper understanding of this methodology to create concepts, see Collier and Mahon, 1993.

2 Durant et al., 2004; Commission on Global Governance, 1995.

3 Kettl, 2002; Knopman and Fleschner, 1999.

this task. This aim implies a conceptual redefinition of governance design and scope. In order to create a radial concept, the analysis in this chapter allows us to widen the scope of climate-change governance by adding in the global factor.

In the second section, the main assumption is that the concept of governance derives from a liberal tradition; in relation to global environmental problems it would have to include the social commitment to preserve and provide global public goods. In this sense, climate global governance refers to the way the provision of public global goods is addressed. This argument opens the door for the third section of this chapter, where I provide a detailed explanation of what kind of global public good clean air is. This allows us to explain on a scientific basis how air pollutants and GHGs impact the climate, and how global and accumulative is the problem. This section concludes by exploring the main sources of GHG emissions. The analysis of energy production and consumption occupies almost all of this section.

The Concept of Governance

The concept of governance arose under the theoretical approach of Liberalism, but, as mentioned before, it has gone beyond some of its traditional principles such as state-to-state cooperation and formal integration. The concept of governance can be traced back to the break with state-dominated social and economic models from 1950 to 1970–80.[4] Theoretically, governance appeared as contesting the International Relations theory of Realism and created debates or complemented some of the Liberal theories at the time. Some theories on which governance builds on are neofuctionalism, transnationalism, interdependence, regimes and neoinstitutionalism.[5]

Building on a multiplicity of liberal theories, a wide range of authors redefined the concepts of authority and power, integrating them to the concept of governance. There was no longer room for the old theoretical debate, where "Realism was long on structure but short on process, domestic politics, and ideas,"[6] and where Classical Liberalism forgot the emerging actors within emerging agendas. New problems started to catch the attention of the international community; there were problems that needed a collective solution with actors that were different from or complementary to the state and could link domestic politics with global issues such as environmental protection or humanitarian aid.

Since then, a set of definitions of governance has dealt with this issue. Some of these definitions have explored how power and authority are exercised in the policy process, usually in an international–domestic relationship. In general, international institutions such as the World Bank, the United Nations Development

4 Weiss, 2000:796.

5 For a brief review of these theories see Viotti and Kauppi, 2010.

6 Keohane, 2002:6.

Program, the Organization for Cooperation and Economic Development, and some national institutions have followed this path.[7] Also, the United Nations Framework Convention on Climate Change (UNFCCC) of 1992—the basis for climate-change negotiations and the Kyoto Protocol—followed this logic. It was not until 1995 that the United Nations coined its own definition of governance in a document called *Our Global Neighborhood.* This report was designed under the newly created United Nations Commission on Global Governance and defined governance as the sum of core values in which individuals and institutions (public and private) manage their common affairs to cooperate or solve conflicts.[8] For the majority of international organizations, governance has been understood as the institutionalization of the international environmental agenda by the United Nations, international corporations and states.[9] Additionally, the concept of governance has been linked to a political neoliberal program expressed as corporate governance aided by a minimal state, where state regulation is replaced by public–private partnerships and market mechanisms.[10]

Authors such as Elke Krahmann argue that this phenomenon appeared in the 1970s in Europe, when the European Union emerged as an alternative political authority that freed the national bureaucracies of certain duties. The "privatization" of the provision of public goods occurred when in the 1990s the International Monetary Fund started to make aid to national governments conditional and categorized countries into the moral parameters of good or bad governance practices.[11] In both cases, the welfare state was under question due to its bureaucracy. This concept of governance was preceded by new-right ideologies in the 1980s, especially in the US, when it appeared necessary to introduce market principles into public administrative systems and business management into the public sector. Later, new trends of privatization started to develop worldwide.[12]

If one considers governance to be characterized as mentioned above, it is evident that these definitions have been deeply contested. One of the major criticisms of the concept is ideological. Governance has been viewed as a global political program related to "good" or "bad" governance practices. This situation led some authors to call this moral qualification a mechanism of domination of the Western ideology, legitimized globally. Some authors call it "governmentality."[13]

7 Weiss, 2000; Douglas, 1999.

8 Commission on Global Governance, 1995.

9 Mingst, 2004.

10 For details on these concepts, see Zürn, 2010; Rhodes, 1996; Mittelman, 1996.

11 Krahmann, 2003:326.

12 Privatization would also include outsourcing, co-production and public–private partnerships. Business management includes also competitive tendering, performance incentives, and internal auditing within a decentralized and customer-oriented structure for increasing accountability, transparency and civil-society participation. Krahmann, 2003:327.

13 As mentioned in the Introduction, the original term "governmentality" was coined by Michel Foucault in 1975. However, some authors have taken this concept and

Expanding the concept to neoliberal institutions, governmentality would imply a moral judgment of the conditions imposed by world leaders to standardize political practices and regularize frequencies, timing and norms of states.[14] Good governance would assume that the goal is to regulate the conditioned interaction between states, markets and civil society, as would happen in developed economies.[15] Nevertheless, this kind of governance can be "a form of violence disguised in the promotion of security"[16] and other global-agenda topics.

However, some other authors explain that good governance was an important component of globalization aimed at the encouragement of representation, democracy and market efficiency at national levels, yet conditioned by international institutions such as the World Bank, the International Monetary Fund and at a certain point the United Nations.[17] In this sense, governance linked two levels of authority (international and national), based on the formal relations between states and markets, where state institutions should guarantee accountability and effectiveness of policies for the market to operate. Philip Cerny suggests that globalization has impacted governance by altering the deeper structures that underlie governance processes and mechanisms.[18] This argument represents the ontologically claimed universalism of the concept of governance.[19] The "universalization" of global governance policies has been criticized because, in this context, power itself inevitably becomes more diffuse, diffracted through an increasingly complex, prismatic structure of socio-economic forces and levels of governance.[20]

As authority is diffused, accountability can be diffused. This leads to another critique that deals with the lack of mechanisms of social participation or the deficit in implementation mechanisms. Some other analysts claim that the sovereignty of nation-states, although contested in the twenty-first century, helps governments to agree on binding conventions designed by global governance institutions.[21] In this sense, governance would have to be, again, an approach focused on the state relating differently with other kind of actors. In other words, it is necessary to "bring the state back in."[22] The state would be the only agent able to guarantee social participation.

A counter-argument comes from those who prefer to see governance as social networking outside the state and apart from international institutions.[23] This

---

contextualized it within globalization. See Sending and Neumann, 2007; Dean, 1999; Larner and Walters, 2004.

14 Zanotti, 2005.

15 Elsenhans, 2001:34–5.

16 Douglas, 1999:135–8.

17 Weiss, 2000; Boas et al., 2005.

18 Cerny, 1999:188.

19 Mol, 2002; Litfin, 2000.

20 Cerny, 1999:190.

21 Jagers and Stripple, 2003.

22 Some authors explain that we can conceive not only one form of state, but different ones with different relations with other actors. For further details, see Evans et al., 1985.

23 Kooiman, 2000.

approach proposes a relation among informal interactions within civil society to organize the public realm.[24] In this sense, social networking or the "privatization" of authority led by civil society is the instrument for contesting the state.[25]

As presented here, the concept of governance can mean different things, can be addressed by different approaches and performed by different actors. In the following section, I present some elements of the different concepts of governance as defined above to help us to develop an original conceptual construction.

Governance as a Policy Process

Robert Keohane and Joseph Nye, contributors in the development of liberal theories, such as Complex Interdependence, established that governance processes no longer fit into the concept of interdependence. These processes have to be inscribed into an era of increasing globalism, with high density of networking, institutional velocity and transnational participation.[26] In this sense, the concept of interdependence evolved into governance, elaborating on several other ideas, such as transnationalism, regimes and institutionalism. Governance can then be understood as "the process and institutions, both formal and informal, that guide and restrain the collective activities of a group within increased globalism."[27] In other words, there is a set of rules of the game in which diverse actors play, aided by institutions.[28]

In this sense, governance can be formal or informal, this dichotomy being the first element in the debate to construct an original concept of governance. The dichotomy formal–informal refers to the emerging groups that could create agreements within legal institutions or within simple social practices.[29] "Informality" means that certain actors can organize to pursue a common goal. The results of this interaction are rules based on implicit understandings and not accessible through written documents or necessarily sanctioned through formal positions.[30] Informal agreements are enforced by the same organizing actors.[31] In the international arena, forms of governance without the presence of formal institutions or the state appear constantly. This phenomenon has created diverse patterns of interaction and responses to authority and government.[32] In this sense, Hart and Pakrash propose

---

24 Boas et al., 2005; Sikkink, 2003:317.

25 Biersteker and Hall, 2002; McAdam et al., 2001. See also Ayres and Macdonald, 2009.

26 Keohane and Nye, 2000:2–9.

27 Ibid.:12.

28 Hart and Pakrash, 1999:2.

29 Keohane and Nye, 2000; Rosenau, 1992:7–9.

30 Boesen, 2006:3.

31 Farrell and Héritier, 2003.

32 Boas et al., 2005:4.

that institutions may operate successfully without formal organizations and one should not assume a one-to-one correspondence between them.[33]

"Formality" refers to explicit rulings generally expressed in institutions, such as agreements or treaties—usually with states as the main performers.[34] The main difference in this kind of relations is that "informal governance-mechanisms are based on trust in persons, rather than on a formal contract."[35] Formal rulings define certain goals and compliance mechanisms for actors (public or private), giving them a ruling framework to interact and to solve conflicts.

The difference between formality and informality relies on the possibility that rules can be enforced either through official sanctioned channels or outside them.[36] Following this argument, governance can be defined in relation to how the formal and informal rules interact and are managed and enforced—in other words, how power and authority are exercised within this interaction.[37] However, "rules on a piece of paper do not constitute social institutions or particular promises."[38] In this sense, the logic of the formal–informal dichotomy needs to be tested by effectiveness in pursuing goals. Thus effectiveness would constitute the second element in constructing the concept of governance as a policy process.

For formal rules, measuring effectiveness does not present great challenges. Compliance with the agreed rules and goals would be enough to incorporate a revision, reform or change to the policy process.[39] This process is performed and controlled under the policy's own rules. However, for areas such as the environment, effectiveness can be measured in terms of policy outcomes or environmental results—especially in informal relations, where usually there is no policy outcome but social practices. Because informal rulings are not clear to everyone within or outside the collectivity, and rely on incomplete information, it is difficult to check on their degree of compliance. Thus one can see only if they have the expected results or not. Some authors propose several characteristics that limit or enforce effectiveness in compliance with formal and informal relations, as shown in Table 1.1.

Informal governance would imply that there is no formal authority to restrain relations, yet effectiveness can be achieved. In fact, scholars such as Ronald Coase have proposed that this kind of informal governance is more effective than when formal authorities (e.g. governments) interfere.[40] In the next section, I discuss this approach for the provision of global public goods. For the moment, the focus

33 Hart and Pakrash, 1999:2.
34 Boas et al., 2005:4.
35 Boesen, 2006:3.
36 Helmke and Levitsky, 2004.
37 Boesen, 2006.
38 Hasenclever et al., 1997:20.
39 Hall and Thelen, 2009.
40 Coase, 1993

Table 1.1 Formal and informal rules

Informal (relation-based)	Formal (rules-based)
1. Relying on private and local information 2. Complete enforcement possible 3. Implicit and non-verifiable agreements 4. Person-specific and non-transferable contracts 5. High entry and exit barriers 6. Requiring minimum social order 7. Low fixed costs to set up the system 8. High and increasing marginal costs to maintain 9. Effective in small and emerging economies	1. Relying on public information 2. Enforcing a subset of observable agreements 3. Explicit and third-party verifiable agreements 4. Public and transferable contracts 5. Low entry and exit barriers 6. Requiring well-developed legal infrastructure 7. High fixed costs to set up the system 8. Low and decreasing marginal costs to maintain 9. Effective in large and advanced economies

Source: Author's elaboration building on ideas of Boesen 2006; Li et al., 2003.

is on the effectiveness of both systems of rules and their linkage, based on how compatible their goals can be between formal and informal sets of norms.

In this sense, governance can indicate the intersection of different forms of relations between different authorities, resulting in a range of transnational structures. In general, formal and informal norms can relate in a cooperative or conflictive way. Some authors propose a four-way scheme in which these sets of norms can relate: complementation, accommodation, substitution and competition (see Table 1.2).[41]

Table 1.2 Relations between formal and informal norms

	Effective governance	Ineffective governance
Compatible goals between informal and formal norms	Complementary	Substitutive
Conflicting goals between informal and formal norms	Accommodative	Competitive

Source: Author's elaboration based on Helmke and Levitsky, 2004; Elazar, 1987.

41 Helmke and Levitsky, 2004:9.

A *complementary* interconnection would mean that informal mechanisms would fill in the gap of the formal ones and vice versa. Processes such as decentralization between orders of government, or coordination among international organizations, NGOs and society are examples of this relation. In an *accommodative* relation, informal norms could be viewed as

> a "second best" strategy for actors who dislike outcomes generated by the formal rules but are unable to change or openly break those rules. They violate the spirit, but not the letter of the formal rules, generally by tempering or modifying the *effects* of those rules. In so doing, they often help reconcile key actors' interests with existing formal institutional arrangements. Hence, although accommodating informal institutions may not be efficient or performance enhancing, they at times ... enhance the stability or sustainability of formal institutions by dampening demands for change.[42]

In other words, accommodative relations can solve the problem of implementation and overlapping by concurrent solutions. However, sometimes these problems cannot be addressed and competition appears. A *competing* relation implies that, to comply with one rule, actors must violate another. *Substitutive* relations are useful when informal relations are created by actors who seek to achieve results that formal institutions were expected, but they have failed, to generate.[43] In all of these relations, the key to effectiveness is the weakness of informal norms. If the informal system of rules is strong, effectiveness is achievable and vice versa.

At this point it is useful to explore the third element of the discussion about formality and informality, the role of institutions and norms. According to Douglass North, institutions are structures that constrain behavior.[44] To complement this definition, Peter Hall and Katheleen Thelen tell us that institutions are "a set of regularized practices with a rule-like quality in the sense that the actors expect the practices to be observed, and which in some but not all cases are supported by formal sanctions."[45] In this sense, institutions can be considered as instruments to operate governance norms.[46] To be useful to governance, institutions need to be effective and dynamic. This can be achieved by exploring the concept of resilience, understood as the degree of adaptation capacity of institutions in the face of changes—internal, external, of leadership, of goals and so on. In other words, how robust and solid are institutions in relation to time and change?[47] According to Peter Hall and Katheleen Thelen, there are three possible ways in which institutions may

42 Helmke and Levitsky, 2004:10.
43 Helmke and Levitsky, 2004:10–11.
44 North,1990.
45 Hall and Thelen, 2009.
46 Hart and Pakrash, 1999.
47 Hasenclever et al., 1997.

change: reform, defection and reinterpretation.[48] Reform involves a change in the main rulings with the support of members. The reform of governance institutions needs to express, in a flexible manner, the changing number of actors and the degrees of attachment to common values in the medium and long run.[49] On the contrary, defection would require dismantling the institution and building a new one. Reinterpretation means that the institutions' actors change principles and beliefs but without dismantling the institution. In this sense, institutionalization can be defined as how values are allocated in specific contexts.[50]

The role of norms and rules is another important element in the concept of governance. Norms that can be widely accepted among actors in institutions are one of the bases for the deep interconnection process.[51] In this sense, institutions are the grantees of norms and act as bargaining forums, as in international environmental institutions.[52] In climate-change topics, this bargaining has had a cornerstone: scientific knowledge. Science has provided a legitimate negotiation table so that actors could debate norms within environmental institutions, whether global, transnational, national, regional or local. However, there seems to be an inconsistent logic between the legitimacy of global agenda issues—based on knowledge and science—with the real implementation of norms—based on policy and politics.[53] For example, for climate change, there was scientific consensus about the anthropogenic dimension of the problem;[54] yet the implementation of these norms was less than successful due to a diversity of business and political groups against it, a topic addressed for the North American case in the following chapters.

The fourth element to construct a different concept of governance under the formal–informal discussion is accountability. For the formal norms and institutions, it is easier to verify behavior regarding budgets, project design and results. There is a clear leader of the institution or governance scheme that can be held responsible. In addition, there are different levels in the policy process that can also be accountable. On the contrary, in informal policy relations there is ambiguity in many areas of the process, yet, as Boesen and Coase suggest, there can be a complete enforcement of agreements. Nonetheless, there are no verifiable mechanisms for rule enforcement or for detection of non-compliance. If informal institutions tried to make the state or other actors accountable, the task would be very difficult because the relations within the policy process usually are personal

48 Hall and Thelen, 2009.

49 Keohane and Ostrom, 1995.

50 Clark, 2002; Myint, 2007.

51 Keohane and Nye, 2000:19; Rosenau, 1992:4.

52 Haas, 1992.

53 Litfin, 2000.

54 Consensus over the "anthropogenic" character of climate change has not been fully achieved. Some scientists still claim that climate change is a natural geological episode in Earth's history. However, when the IPCC published the three scientific reports in 2006, consensus was almost achieved among the epistemic community.

and not confirmed by anyone else. To be clear, there would be no legal mechanisms to make the authority accountable, but there might be political ones. Guillermo O'Donnell points out that informal mechanisms usually cohabit with formal ones in developing societies,[55] the first being the ones that dominate. He situates the policy process in a "gray area," where both norms coexist.[56] In other words, the interaction between formal and informal institutions may actively reconfigure competing norms into more compatible ones.[57] However, the necessary condition for interactions (formal–informal) needs to be law enforcement. This would imply that the formal part of it should be dominant in the policy process.

Relocation of Authority: A Diversity of Actors

In order to continue with the discussion of how formal and informal sets of norms interrelate in governance processes, the role of authority needs to be explored; this therefore constitutes the fifth element. It is "through the exercise of authority that decisions are made and implemented and the coherence of collectivities thereby preserved."[58] In a Weberian concept of authority, authority is the power accepted as legitimate by others subjected to it. This means that an important part for authority deals with legitimacy.

> Although the rules and procedures for exercising control may be codified and cited as the "authority" for policies pursued, this is only a shorthand and legalistic way of referring to complex relational processes ... authority is maintained only to the degree that the membership continues to treat the leadership as having the right to lead [where] legitimacy is the acceptance attached to the actions by the membership.[59]

Some authors see authority as a function of governance necessary to establish and implement collectively binding decisions.[60] The relation authority–governance drives in two directions. On the one hand, there are those like James Rosenau who argue that authority is a system of rules that works only if accepted within

55 Furthermore, informal ties are also present in developed countries through technological advances, sometimes presenting the same accountability problems.

56 Guillermo O'Donnell frames this assertion in an analysis of democracies. He proposes that some democracies do not follow in the traditional category of representative democracy—especially in the region of Latin America. He calls this new categorization "delegative democracies," which agree with elements of traditional authoritarian practices but combine them with elements of transition to representative and liberal democracies. See O'Donnell, 1993.

57 Helmke and Levitsky, 2004:13.

58 Rosenau, 1990:186.

59 Rosenau, 1990:190.

60 Draude, 2007.

governance processes and without any central authority.[61] Krahmann agrees with Rosenau and defines governance as the structures and processes that enable governmental and non-governmental actors to coordinate their interdependent needs and interests through the making and implementation of policies in the absence of a unifying political authority.[62] On the other hand, there are those who assert that a central authority is needed to perform governance, either in the form of hegemony or of a paymaster.[63]

This research explores the second route, where authority is needed to lead institutions to accomplish norms, especially if thinking of policy creation and implementation through transregional climate approaches. In this sense, the leadership of the paymaster in formal schemes would normally imply that authority rests in states, and it is performed by governments.[64] In informal schemes, this leadership can be exercised by anyone. However, even in non-hierarchical regulation processes "the shadow of [state] hierarchy" is present.[65] Also, this needs to be taken into account to provide global public goods, where having a paymaster reduces the free-riding phenomenon because one actor has the legitimate authority to punish defectors and absorb the majority of transaction costs.

In this scenario of diffused authority among different actors in the governance process, who could perform as paymasters? As mentioned before, some scholars argue for not leaving the state aside in governance processes.[66] Their main argument is that the state has authoritative capacity to enforce rules and thus to give stability to transnational complexity, where there is overlapping of interests, levels, structures and actors.[67] For others, governance can be understood as a complex set of governing structures and processes that goes beyond governments of states.[68] In this approach, formal relations would include governments but not as centralizers of authority. In other words, governance includes governments, but as another channel for commands to reach goals and policies.[69]

This diffusion of authority and the debate about the role of states has resulted in the emergence of different ranges of authority, legitimacy and influence, with local, national, transnational or global scope. Groups in civil society such as non-governmental organizations, local, regional and global governance institutions, and firms have taken advantage of this situation. The diversity of actors interacting in a

61 Rosenau, 1992:2–9.

62 Krahmann, 2003:330.

63 The proposal that a regime would need to have one member to become hegemonic was discussed by Stephen Krasner in his theory of regimes. Oran Young elaborated on that idea, and proposed three categories of leadership: negotiated, imposed and spontaneous. In the imposed and negotiated, a leader would be present. See Krasner, 1983 and Young, 1992.

64 Weiss, 2000:801.

65 Draude, 2007.

66 Draude, 2007:7; Cerny, 1999:197.

67 Cerny, 1999:197.

68 Weiss, 2000:795.

69 Rosenau, 1992:2–4.

governance process can be immense, and the kind of actors can be heterogeneous.[70] As Elinor Ostrom suggests, the number or type of actors is not important; rather, what matters is the strength of the understandings of common interests, the way they share the costs of cooperation, and their assets (information, decision-making structures and authority).[71]

Among the multiplicity of actors, the role of the state in governance processes is still important, given that it can change or influence the general frameworks to address global public goods. Thomas Weiss establishes that the link that relates "governance patterns with national and global levels lies in solving the collective action puzzle to provide public goods."[72] To provide these types of goods, some scholars minimize the importance of the state;[73] others argue for transforming the state into a competitor[74] or even into an efficient, innovative and profitable corporation.[75]

Hart and Pakrash have suggested that the state needs to rearticulate itself to continue providing and redistributing services and goods. One of the ways to do this would be "to co-produce collective goods with their citizens [and] to include civil society in institutional design and implementation."[76] Thus the state would obtain legitimacy by redistributing gains for winners and losers. An interesting view is that of Evans, who notes that either states become more repressive in providing collective goods or they start, as mentioned above, to include society in the decision making.[77] However, if we understand governance as a multi-layered policy process, the separation between public and private is no longer clear due to the fact that, nowadays,

> public authorities engage directly in economic activities through state-owned or controlled companies and private firms take on public functions such as setting standards or providing health care. The rise of the "competition state" and the dramatic expansion of the social responsibilities of business firms have blurred the once clear line between public and private sectors.[78]

This mix in agency is explored in the next section, where public and private spheres find intermediate areas, especially for climate-change topics. Nonetheless, for governance to be operable, public authority is needed for certain aims: (1) to internalize, regulate and implement global governance issues and agreements;

---

70 Keohane and Ostrom, 1995:6.
71 Ostrom, 1990; Keohane and Ostrom, 1995:7.
72 Weiss 2000:807.
73 Ibid.
74 Cerny, 2006:379.
75 Turnbull, 2002.
76 Hart and Pakrash, 1999:15–16.
77 Evans, 1997.
78 Kobrin, 2008:4.

(2) to design the mix of policy mechanisms and coordinate them by using budget resources; (3) to correct market failures in providing public goods (national or global); and (4) to create incentives for other actors through policy and sanctions to accomplish goals.

The discussion about the state's role in governance processes leads us to the debate about the scope of governance. How wide is it, and how wide should it be? Some authors see governance as a process of coordination between levels of formal authority: global, multinational (regional), national and local (or sub-national).[79] To talk about levels introduces the idea of hierarchy, where the global level is above the multinational (regional) and the national levels. To talk about levels also means that there is a separation between global/regional/national agendas, objectives and goals. Yet, for environmental protection and climate change this process of differentiation is not always as clear as it may seem.

Isidro Morales explains that a multi-layered governance approach is more accurate in explaining new power relations within a globalization framework among states, markets, institutions and non-government actors. Layers can be conceived as spheres of authority well differentiated, in spite of being interwined with the state.[80] This vision encompasses more actors and goes beyond the hierarchy of the level approach,[81] transcending formal links. It also allows the construction of governance institutional structures that belong not only to one of the levels but overlap in a transnational form. Finally, this approach clarifies the confusion between governance and government. Government represents a central public authority under only a national level, while governance represents a fragmented combination of public and private actors at different levels, as well as formal and informal norms regulated by different types of authorities.[82] To further clarify the relation between governance and government, Fabio Masini suggests that to speak of multi-layers implies addressing institutional governance structures that can be composed of levels of cooperation,[83] in other words levels of government.

As Thomas Weiss puts it, "[g]lobal governance refers to collective efforts to identify, understand, or address worldwide problems that go beyond the capacities of individual states to provide government-like services in the absence of world government. Global governance encompasses an extremely wide variety of cooperative problem-solving arrangements."[84] In this sense, global governance

79 Krahmann, 2003.

80 Morales, 2008:20–21.

81 Lisbet Hooghe and Gary Marks (2003) use the term "multi- or polycentered" governance to describe what in this thesis is called multi-layer governance. This also reflects the fact that governance processes do not necessarily have hierarchical authority.

82 Krahmann, 2003:326–30.

83 Masini, 2011.

84 Weiss, 2009; see also: Amen et al., 2011.

is the alternative to the false dichotomy between a mere interstate system (intergovernmentalism) and a world state (supranationalism).[85]

Relocation of authority is directly connected to institutions, processes and actors. The next section presents a new interpretation of these elements to integrate a different concept of governance useful for analyzing climate change from a transregional perspective.

A Redefined Concept of Governance

Elaborating on the previous elements and definitions, this section explores a new way of conceiving governance. In my view, governance can be understood as processes of coordination among different actors and government levels in diverse constructed formal and informal relations expressed in multiple layers, with a leader granted legitimate authority. In the case of environmental governance, the purpose is generally to provide public goods.

This proposed definition has elements worth exploring. In the first place, coordination will be conceived as a relational process that performs in different ways. Coordination can occur within layers, and among levels or actors; it can occur also among different layers of authority for different topics. These spheres of authority, although having frontiers (sectors, issues, problems), are flexible enough to change and incorporate other issues and actors. This characteristic is achieved through cooperation and operated through regulation.

Multi-layer coordination normally happens within a transnational scope, where the relational process is in the hands of governance institutions, but the implementation of policies needs governmental appropriation. In this sense, governance institutional structures and government operations cannot be completely separated. To make governance operable, centrality at certain layers needs to be present. This would be true if we defined governance not as lacking public authority, but as one expression of it. This does not mean that international organizations or national governments would be ignored. In fact, "states could reinforce their governance capabilities by delegating authority and power to sub-nation units, transnational institutions, and/or supranational ones."[86] States could also reinforce their capabilities by taking advantage of other actors' trends, norms and rules to extend them on a national basis or incorporate them into national policies. As Kal Raustiala notes, global governance can reinforce the implementation capacities of governments in topics such as environment.[87] This is true for the case of climate-change governance in North America, where

85 Enderlein et al., 2010.

86 Morales, 2008:22.

87 Raustiala, 1999.

"global trends, despite operating in the supra-territoriality of flows, concentrate and operate within territorialized and delimited spaces."[88]

In this coordination scheme, Jan Aart Scholte suggests that regulation in a post-sovereign world has mainly occurred in a bureaucratic way[89] and tends to facilitate governance processes. However, diverse processes coexist within multi-layer governance, one that represents the status quo and situates the state as the main regulator, and another that presents a variety of actors that can play the role of regulators. Scholte notes that "the irruption of the global era does not mean that everything is moving and changing. There are in fact various elements of continuity from the 'old' order and the new emerging one."[90] In other words, the flexibility of bureaucratic principles enables continuity and change to coexist.

Within the continuity-and-change discussion, multi-layer governance also implies that traditional territorial approaches cohabit with new forms of institutionalizing territorial regulation and governance. One example is regionalization trends within a multi-layer framework, where traditional territories within the state become territories beyond state borders or where they are transformed into virtual regions. Associations of communities (either neighboring or not), transnational institutions or telecommunication interconnections facilitate this process.

Traditionally, the concept of coordination would be inscribed into federalist theoretical frameworks.[91] Coordination would employ the federalist principle of "shared and self rule," usually expressed by levels of government where each one has a certain autonomy in policy making. However, in a multi-layered framework, coordination happens when one issue needs to be dealt by different orders of government and other actors. For example, there are several forms of regulating coordination.[92]

These forms of coordination are possible because of the different scales of regulation. Regulation can take several forms. The first is *convergence.* Convergence is understood as the trend of policies to look alike regarding their structures, processes and performance. The drivers of this convergence can be ideological, economic, external (contextual) or internal (within the actor or layer).[93] The second is *harmonization*, defined simply as standardization of policies or processes generally throughout specific institutions;[94] the third is *linkage*, which

88 Morales, 2008:16.

89 Scholte, 2000: 157.

90 Ibid.

91 The theory of federalism states that there can be different types of relations among federated units and national governments. This usually implies a clear hierarchy in policy making. Many scholars have discussed federalism; among the most relevant in terms of theoretical debate are Elazar, 1987; Kincaid, 1998; Brown, 2002; Watts, 1999.

92 Brown, 2002.

93 Drezner, 2001.

94 Tuerk et al., 2009:343.

deals with establishing certain instruments for linking governance processes without an institutional building.[95]

In a regulatory scheme, governance can be performed at subsidiary layers, especially for providing public goods. John Ferejohn and Barry Weingst state that in order to decide about subsidiarity it is necessary to consider which jurisdiction will be more capable of providing public goods. However, this depends on the characteristics of the topic to be addressed as well as the level of enforcement necessary.[96]

In sum, global governance has been defined in several ways and with various scopes. The majority of the definitions include the same elements. They all address the fact of relocation of authority, of broadening the spectrum of actors involved in policy processes, of a variety of scopes that tend to erase the borders between static levels of interaction. There is no longer any outside or inside; layers are interconnected, especially for environmental topics. Other essential elements are institutions and norms as the main drivers to regulate processes towards the provision of global public goods.

Chapter 2 explores the characteristics of these types of goods and open up the discussion about what type of actors, norms and process are better suited to address climate change and to ensure clean air as a result of policy processes.

95 World Business Council for Sustainable Development, 2007.

96 Ferejohn and Weignst, 1997:xiii.

Chapter 2

Climate Governance: Global Public Goods and Science

Introduction

The previous chapter explained the importance of clarifying the concept of governance and proposed several essential elements. One of the most important goals of governance processes at a global scale is the provision of public goods and the management of public "bads." The first section of this chapter discusses what type of public goods can be provided under climate governance. Clearly, this discussion needs to be framed within the scientific approach to the problem. Science has been the basis for creating a variety of responses to the phenomenon and can explain why the problem of climate change is in essence global. The second section addresses the scientific arguments for designing climate governance. It briefly explains how the climate system works and what the greenhouse gases are. In addition, it explores how these gases behave in the face of human activities.

Global Public Goods and Climate Governance

When we talk about global governance it sometimes seems that concepts vanish into thin air. In the case of climate change, we are indeed. The material object for climate governance is air and the policies for cleaning it. In other words, the main problem with climate change is that air (or the atmosphere) has been filled up with pollutants that absorb and retain heat. Heat produces changes in temperature and precipitation, affecting climate. In this sense, cleaning the air (and consequently tackling climate change) needs to be considered a global public good, not an environmental private commodity.

This section defines global public goods and establishes the difference between these and private ones. It starts with a detailed discussion of the three constitutive elements of the concept: *globalism*, *publicness* and what is understood as a public *good* (or *bad*). The section continues by analyzing how, in environmental issues, instead of public goods we generally find public "bads" (or externalities) in need of correction. These externalities are inscribed into market failures, due, in many instances, to incomplete markets (which is also a characteristic of environmental issues). This leads to identification of who can provide the corrections and prevent the externalities, especially for the case of climate change.

Clean Air and Climate Change: Three Elements of Global Public Goods

We need to distinguish between public and private goods. The classical definition of public goods establishes that they are non-rival and non-exclusive.[1] "Non-rivalry" refers to the fact that one person does not compete with another for the use of the public good, because the first person's use does not reduce that of another. "Non-exclusivity" refers to the fact that no one can be excluded from using the good, as this would be extremely costly or technically impossible.[2] For example, public goods can be (1) tangible, as roads, infrastructure, clean air; (2) intangible, as knowledge, security, peace; and (3) policy outcomes, such as environmental protection.[3] In contrast, private goods are excludable in benefits or costs and rival in consumption.[4] The reason for this is that private goods have clear property rights and can participate easily in the market. In other words, this characteristic gives markets the ability to allocate resources efficiently.[5] For public goods, especially environmental ones, the market is unable to allocate resources in efficient ways. Because of this, market failures appear: the market lacks elements, and is thus incomplete.[6] This means that, in the process of production of a private good, private producers pay the costs, but a divergence between private costs and social costs occurs as society as a whole pays for the damage. An example of this is pollution resulting from industrial processes.

As established earlier, market failures occur when markets are incomplete, lacking the clear property rights necessary for exchange.[7] This situation has obliged governments to correct these market failures and act as grantors of property rights over these assets.[8] A counter-argument can be found in the Coase Theorem developed in 1960. Ronald Coase pointed out that private partners would reach a Pareto- efficient[9] solution regardless of the party to whom property rights have been assigned (assuming there were no transaction costs).[10] However, this assumption does not take into account that transaction costs exist and have to be paid by someone.[11] "The Tragedy of the Commons," by Garrett Hardin, supports

1 For this classical definition, see Samuelson, 1954.

2 Hanley et al., 1997:37.

3 Kaul et al., 2003:15–17.

4 Kaul and Mendoza, 2003:79.

5 Kahn, 2005:14.

6 Hanley et al., 1997:22

7 Hanley et al., 1997:24.

8 Ibid.:25.

9 An allocation is Pareto efficient or optimal when it is impossible to relocate resources to make one person in the economy better off without making someone else worse off. See ibid.:24.

10 Coase, 1993:116–19.

11 Transaction costs are the costs of devising, monitoring and enforcing rules to meet a particular objective. See Keohane and Ostrom, 1995:2.

this premise by arguing that "freedom in commons brings ruin to all."[12] In this case, society as a whole would pay the cost of private actions or bargaining. In other words, public goods are subject to underprovision due to open access.[13]

Coase recognized that there were limitations to his theory. First, although the main premise is that the market would resolve disputes, he accepted that "there is no reason why, on occasion, such governmental administrative regulation should not lead to an improvement in economic efficiency."[14] In this sense, Hardin suggests that the tragedy of the commons could be prevented by coercive laws or taxing devices that make it cheaper for the polluter to treat these pollutants than not to treat them.[15] But there is a possibility that transaction costs might be so high that the existing law would not be changed to allocate property rights.[16] Elinor Ostrom points out that sometimes government investment has disrupted local institutions and practices, which has caused incentives for cooperation to vanish.[17]

However, if we think of public goods as environmental assets, there would be an enormous social cost if property rights were only privately allocated. Due to this, in general, environmental property rights are in the hands of authorities or in a combination of the public and the private domains. Examples of this situation can be found all over the world, but specifically in the three countries of North America where environmental assets are owned either completely or partially by local authorities, as in Canada and the US, or by federal ones, as in Mexico. Environmental assets in some localities are managed in a communitarian manner, dividing the property rights between private individuals, a collectivity (the community) and the local authority. Finally, although the process for proving a global public good (or transforming a bad one into a better one) could be priced clearly, the outcome (public good) would not be owned by anybody. An example of this characteristic is clean air, where the transaction costs of transforming polluted air into clean air are high and not paid equally by all. Then the problem of free-riding appears, as there are no incentives to reduce pollution, and no visible individual costs or benefits from doing so.[18] For global public goods, dealing with free-riding at all scales (global, regional etc.) has challenged the task of producing and providing them.

The second difference between public and private goods is that in the latter it is known how consumers behave, because when they pay for the good, they are already expressing preferences. For public goods, non-excludability makes it difficult for consumers to reveal their preferences. Because of this, imperfect

12 Hardin, 1993:9.
13 Keohane and Ostrom, 1995:4.
14 Coase, 1993:122.
15 Hardin, 1993:11.
16 Coase, 1993:123.
17 Ostrom, 1995:125–7.
18 Stern, 2007:27; Kaul et al., 2003:ii–iii; Hanley et al., 1997:43.

information arises, forming obstacles to correct pricing mechanisms.[19] For example, markets of energy, land use or water do not reflect the consequences of different consumption and investment choices for the climate.[20] Non-exclusivity, non-rivalry, incomplete markets and uncertain prices generally apply to pure public goods.[21] However, there are certain cases where exclusion and rivalry can be imposed on them as a policy decision. An example of exclusion of this type would be when there are charges (fixed pricing schemes) for the public good; an example of rivalry occurs when consumers are limited by groupings, so that the public good would reach different people in a differentiated manner.[22] Rivalry would not be a problem for abundant resources, such as oxygen, but probably would be for scarce ones, such as water.

Partial public goods are exemplified by fossil fuels. Fossil fuels are not a global public good but a private one (in the hands of private or public actors). Leaving for later discussion the debate about who can own and exploit these fuels, it is necessary to acknowledge that they should be addressed as partial public goods due to their impacts. In this sense, they can be easily overused or appropriated by private actors with no responsibility for their impact on the commons, which constitutes social free-riding. In a globalizing world "it has become more difficult to exclude non-paying users (free-riders) from outside national boundaries from benefitting from nationally-provided collective goods."[23] The kind of resources that compose the "common natural pool"[24] of humanity are generally in this gray area of quasi-private in origin (or in production) and quasi-public in impact (or in consumption/environmental degradation).[25] How public a good is and what range of action benefits and costs reach need to be addressed in a more precise manner. Generally, these private-in-origin and public-in-impact goods are found in the public domain—not necessarily in the state domain—with no or unclear property rights.[26]

What is "public" about these goods? Their "publicness," as Philip Cerny suggests, is essentially normative: it denotes matters that people think ought to be treated as common or collective concerns.[27] Yet who represents this collectivity? The public would include every person, group, civil organization, business[28] or

19 Kahn, 2005:17; Kaul and Mendoza; 2003:80.

20 Stern, 2007:27.

21 Hanley et al., 1997:42.

22 Kaul et al., 2003:iii.

23 Cerny, 1999:200.

24 The theory of "common-pool resources (CPR)" was drafted by Robert Keohane and Elinor Ostrom. They defined CPR as the natural resources humanity has, but different from a pure global public good in that they are sometimes subject to rivalry or excludability in consumption or impacts.

25 Lipschutz and Fogel, 2002.

26 Kaul et al., 2003:7.

27 Cerny, 1999:199.

28 Business is usually seen as a private actor, but when the sector acts as a public citizen, for example when it participates in the stock exchange, it can be considered part of

family, as well as states and markets.[29] In other words, "public goods are the public's goods"[30]—or everyone's goods.

The public domain depends on two criteria. First, how public and participatory the decision-making process is; second, how equally the benefits are distributed to the public.[31] In the first case, a normative approach is necessary. Phillip Cerny establishes that there are four characteristics to normativity regarding public goods: regulation, production, distribution and redistribution.[32]

The second case deals with human activities throughout history, where two kinds of goods have existed simultaneously: the natural public goods and the human-made public goods, such as infrastructure and security systems. Yet these last ones can impact the natural public goods, creating public "bads."[33] When these public "bads" impact other people differently from the ones who caused the damage, an external damage occurs. Externalities are understood as side-effects or spill-overs (which can be damaging or beneficial) resulting from market transactions, generally not intentional.[34] If the spill-over of one jurisdiction to another is harmful, as in the majority of cases of environmental pollution, there will be an extra cost of preventing others' pollution processes.[35] In externalities or public "bads," such as climate change or polluted air, the damage to one person does not reduce the damage to others, not even to the person who created the pollution in the first place. This characteristic confirms that there is non-excludability and non-rivalry of impacts or of consumption.

A public "bad," such as polluted air and climate change, can also operate as a dynamic externality. This occurs when, for example, air pollutants such as pesticides, persistent organic pollutants (POPs), radioactive wastes or chlorofluorocarbons (CFCs) accumulate over time, causing persistent damage in the future and enormous social costs for the next generations.[36] There is also a need to bear in mind that polluted air and climate change are externalities that bear a burden of uncertainty about potential size, timing, location of impacts and future costs.[37]

For some environmental public bads, such as oceanic depletion, water contamination and air pollution, the spatial scope of publicness is generally wider than for other environmental issues, such as urban solid wastes and pesticides in agriculture. Although some public goods are constructed to be global, such as security, some are essentially global. In this sense, the nature of some public goods

---

the public domain.

29 Hoppkins, 1999.

30 Kaul et al., 2003:14.

31 Kaul and Mendoza, 2003:85.

32 Cerny, 1999:200–203.

33 Kaul et al., 2003:7.

34 Kahn, 2005:21; Baumol and Oates, 1988:17.

35 Kaul et al., 2003:xv.

36 Stern, 2007:28; Kahn, 2005:31.

37 Stern, 2007:28.

is conceived under globalization and its global provision is the main concern. As explained by Kaul, "[i]ndeed, the clamor against globalization can be interpreted as a call for better provision of global public goods."[38] "Globalism" of public goods/bads means that they span different levels and layers of governance schemes all over the world. The recipients of the global public good/bad are basically the world population, transnational actors, non-state and nonprofit actors, and states. In other words, a *global* public good is one whose benefits or impacts extend to all people, countries and future generations;[39] it transcends time and territorial space. Although air circulates freely all over the planet, it is *clean air* understood as a global public good that should be provided by climate governance. Consumption of clean air is not exclusive and benefits are also available to all consumers.

Who Can Produce and Provide Global Public Goods?

Either public authorities or markets can provide public goods.[40] However, for producing *global* public goods, some authors suggest that it is not enough that nations make their own efforts; it is necessary for common research, knowledge and surveillance to be in the hands of global institutions.[41] In other words, the first element in the production of global public goods is an institutionalized political and policy process, "where regulations of private activities [with a global scope] are the primary means to produce some types of public goods."[42] Institutional arrangements need to exist that can create incentives to provide them through such means as linking private with public goods or the activities of political entrepreneurs.[43] Yet it is necessary to be careful with governmental intervention. Sometimes governments, instead of correcting divergences between private and social costs, widen them, causing distortions.[44] The second element is financing the provision of global public goods, which requires an adequate allocation of resources. The last element is management to implement this provision in the best and most efficient manner.[45]

The scale of provision of public goods depends on the range of benefits. In theory, environmental public goods with a narrow range of benefits, such as cleaning lakes or rivers and conserving endemic species, need to be provided locally. Those with nation-wide scope should be provided centrally, and those with global range, globally.[46] To ensure that global public goods are provided, it

38 Kaul et al., 2003:4.
39 Kaul et al., 2003:10–23.
40 Kahn, 2005:18; Kaul et al., 2003:9.
41 Kaul et al., 2003:12.
42 Smith et al., 1997:2.
43 Keohane and Ostrom, 1995:13.
44 Kahn, 2005:20.
45 Kaul et al., 2003:15.
46 Kaul et al., 2003:xiii–xiv.

is assumed that nation-states alone cannot provide them adequately. They require global solutions.[47] Even though the public-goods theory has been state centered, there is a need to redefine institutional interactions and spatial management through policy entrepreneurship to truly understand global public goods.[48]

Some authors claim that, if the benefits and impacts of regional or national provision spill over to the next jurisdiction, then the issue should be treated globally. "The more global is the region over which the benefits extend, the greater is the need for global policy instruments."[49] due that the local, national or regional underprovision of a global public good can be felt everywhere in the world.

However, the solution does not always come from the same arena; global problems are not always efficiently solved by global solutions. For example, although some environmental public goods are in general global, the impacts of not having them are very different and specific according to regional particularities and circumstances.[50] In other words, in spite of their "globalism," responses have territorial and administrative bases that have tried to be coherent, covering the same territory and activities. Another example is when actors in multi-layer governance voluntarily accept global norms and common environmental standards, and adapt them for their territories.

Some of these responses have been acting in regional and national arenas. This offers the opportunity to situate localism not as the antithesis of globalism, but as one dimension of global processes and relations. In these processes new institutional arrangements are created among different actors with different goals based on different power relations, with the participation of a diversity of levels of authority and governance layers.[51] In this sense, the provision of global public goods cannot rely only on governments or markets. If it relied only on markets, this would certainly cause unequal distribution of benefits; if it relied only on governments, it would be difficult to find a cost-effective way to pay for it.[52] Sverker Jagers agrees with this assertion and notes that global environmental public goods should be the object not only of national policies but also of other authorities and actors.[53] The global environmental provision of rules needs to be internalized partially or completely by different actors in different ways.[54] Contemporary states have had the ability to adapt to a world where they are allowed to enter into durable and enforceable ("credible") commitments to the common resolution of

47 Stern, 2007:xiii; Kaul et al., 2003:12; Keohane and Ostrom, 1995:11.

48 Kaul et al., 2003:5–6.

49 Kaul et al., 2003:xiv.

50 Cohen et al., 2004; Kaul et al., 2003.

51 Hocking, 1999; Duchacek, 1988.

52 Ibarrarán and Boyd, 2007; Kaul et al., 2003.

53 Jagers and Stripple, 2003.

54 This "double-level" strategy between global and national institutions has been used by actors in global governance, especially for climate change. See Agrawala and Steinar, 2001.

global public goods.[55] States have been "playing safe" in global ruling systems, especially in the environmental domain. Yet Cerny perceives the need to escape this pernicious dynamic by understanding global environmental governance (especially for climate change) as a "multi-layer" strategy,

> which includes not only firm-to-firm diplomacy, but also transgovernmental networks and policy communities, internationalized market structures, transnational cause groups, and many other linked and interpenetrated markets, hierarchies, and networks.[56]

In other words, for the provision of global environmental public goods, it would be necessary to adopt a multi-layer governance approach that could design organizations and regulations operated by formal institutionalization and that could relocate authority.

Good vs Bad Public Goods: Clean Air and Greenhouse-gas Emissions

Recovery of clean air is the basic global public good at stake. In this sense, the goal of climate governance is to reach stable temperatures and climate patterns, and to avoid uncertainty caused by extreme warming.[57] In addition, climate governance aims to mitigate and manage greenhouse-gas emissions (GHGs) and adapt to their impacts.

As established earlier, greenhouse-gas emissions or air pollution in general are partial global public bads, because they can be regulated to address non-exclusivity.[58] For example, pollution permits or cap-and-trade systems are property rights that basically allow someone to pollute the atmosphere and compensate for that, granting only some sort of exclusivity in its use.

However, polluted air would be essentially non-rival in the sense that pollution permits do not allow air to be owned by someone (characteristic of a private good), but only limit the actors' right to use it in a certain way, in this case to pollute it.[59] When greenhouse gases (public bads) are emitted to the atmosphere, pollution impacts have social costs in the entire world. These impacts depend on the mitigation and adaptation conditions available in different societies. For instance, they would be very costly for poor societies and less so for rich ones, although GHGs were mainly emitted by rich ones.[60] For policy design, the costs

---

55 Cerny, 1999:191.

56 Cerny, 2006:383.

57 According to The Stern Review, stabilizing GHG levels in the air would require maintaining between 450 and 550 PPM CO_2e. The current levels are at 430 PPM and rising 2 PPM each year. So stabilization means bringing down emissions to 80 percent below current levels annually. See Stern, 2007:xvi.

58 Kaul and Mendoza, 2003:83.

59 Ibid.:84.

60 Stern, 2007:31–4.

of emissions, mitigation and adaptation are certainly rival and excludable, making GHGs a *partial* global public bad.

There can be two approaches for policy design to deal with partial global public bads. The first is to focus on GHG sources in order to mitigate them, by paying either for prevention or for pollution. The second is to address their consequences by adaptation. For the choice of prevention, it is necessary to accept that the global public bad (polluted air) came from human-induced GHG emissions in a private and differentiated way. In this sense, the provision of a solution would deal with designing policies to cut private GHG emissions (an accountable and tangible asset) to avoid externalities. GHG emissions usually accumulate in time and can be conceived of as dynamic externalities, as shown in Table 2.1.

Table 2.1 Scale and duration in the atmosphere for selected pollutants

	Hours	Days	Weeks	Months	Years and centuries
Global					CO_2, CH_4, N_2O, SF_6, HFC, PFC, HCFC, CFC
Hemispheric			Tropospheric Ozone	CO_2	
Regional		SO_2, NO_2, NO, NH_3, PM10			
Local	SO_2, NO_2, NO, NH_3, PM10				

Source: Author's elaboration based on the GEO4 Report of the United Nations Environmental Program, 2007.

Any strategy would then need to take into account that "it is longer possible to prevent the climate change that will take place over the next two or three decades, but it is still possible to protect our societies and economies from its impacts to some extent."[61] If we chose to simply pay for the right to pollute, we would only be bargaining for externalities and transferability, forgetting the global social cost. A transferable externality implies that each individual protects himself from external damages "by simply transferring an environmental risk through space to another location or through time to another generation."[62] An example of this was registered when the US Midwestern industrial states reduced regional air pollution

61 Stern, 2007:xvi.

62 Hanley et al., 1997:33.

problems (SO_2 depositions) by building taller stacks for emissions. Air currents transported these emissions to the northeastern states and to eastern Canada.[63] Sometimes, environmental policies that allow unilateral transfers of pollution instead of encouraging cooperative solutions will result in expenditures on self-protection that will be very high.[64] In this sense, the bargaining scheme of Coase would appear complicated, because "the spatial dimensions of the environmental resources link private activities to the public good matter."[65]

When pollution permits are used, governance and authority institutions (usually governmental ones) are driven to take into account this social cost. For example, even with these exchange institutions, environmental damage caused by one person could damage an entire ecosystem. In this sense, it would be very difficult to measure how much polluted air every person makes or transfers per day and what percentage of ecosystemic damage this would represent. This leads to the problem of pricing for damages and the way these costs would be exchanged. The combination with market mechanisms appears to be the best solution.

Whatever mitigation focus a country chooses, "a ton of carbon emitted anywhere on earth has the same effect as a ton emitted anywhere else."[66] This means that everyone would need to adapt to nonexclusivity and nonrivalry of impacts of GHGs. Adaptation can be understood as the "adjustment in natural or human systems in response to actual or expected climatic stimuli or their effects, which moderates harm or exploits beneficial opportunities."[67] There are three approaches to adaptation. First we have the concept of adaptation under the UNFCCC, where there is a need to provide reliable assistance for high-priority implementation projects. The second approach is to integrate adaptation with development, and the third appears as climate insurance to compensate for vulnerability.[68] In general, some sectors are more vulnerable than others, such as health, agriculture, water and biodiversity. In addition, some societies are more vulnerable than others due to their geographical location and low levels of development.[69] In this sense, markets alone would find it difficult to address vulnerability. The social cost of climate change would need to address adaptation in the same way as any other social policy.

Whatever approach (mitigation, adaptation, or both) is taken in the short run, the costs of not addressing this situation correctly will become more and more expensive over time. For Schelling, the costs of reducing emissions of CO_2 would

63 Smith et al., 1997:7; Hanley et al., 1997:33; Dorfman and Dorman, 1993:439.
64 Hanley et al., 1997:36.
65 Smith et al., 1997:8.
66 Schelling, 1993:468.
67 Intergovernmental Panel on Climate Change, 2007b.
68 Burton et al., 2006.
69 United Nations Environmental Program, 2007.

cost hundreds of billions of dollars in perpetuity.[70] A more precise analysis was presented by Nicholas Stern in 2006.[71] In his report, Stern pointed out that,

> if we don't act, the overall costs and risks of climate change will be equivalent to losing at least 5% of global GDP each year, now and forever. If a wider range of risks and impacts is taken into account, the estimates of damage could rise to 20% of GDP or more. In contrast, the cost of action—reducing greenhouse gas emissions to avoid the worst impacts of climate change—can be limited to around 1% of global GDP each year.[72]

The report emphasizes that the world does not need to choose between tackling climate change and promoting growth and development. Instead, addressing climate change can become a "pro-growth strategy"[73] in combination with mitigation and adaptation strategies.[74] In fact, this combination could mean that other actors participate in the provision of the global public good, and become leaders of growth and development; the price to be paid might be to reduce the state's autonomy in return for an increase in problem-solving capacity and legitimacy.[75]

As established in this section, the risks of not taking action on climate change can be very costly over time, both socially and in terms of GDP. In addition, the ecosystemic cost is priceless. The following section discusses these impacts through the lens of science. A social sciences perspective cannot adequately present a deep analysis and defy the science behind climate change. Thus the next section addresses only the scientific facts and the debates important for decision making regarding climate governance and the provision of global public goods.

The Science behind Climate Governance

In any discussion of the science of climate change it is common to hear about the impacts of the changing climate: the melting of the ice caps, unusual storms and off-season hurricanes, longer droughts, long-lasting precipitation and floods are what the public generally understands by climate change.[76] All these events pose a risk to all life forms on the planet. However, this section intends to go one step back and try to explain why these natural disasters happen at a fundamental level. Understanding what greenhouse gases are and how they behave is the basis for

---

70 Schelling, 1993:468.

71 The "Stern Review Report" was launched in 2006, but not published until 2007.

72 Stern, 2007:xv.

73 Ibid.:xvii.

74 Van Kooten, 2004.

75 Börzel and Rise, 2002

76 For a detailed analysis of these events on a regional and national basis, as well as for information on all environmental topics, see United Nations Environmental Program, 2012a.

assessing the importance of these elements for climate governance. In addition, some important data show that the increase of human activities modifies the behavior of these gases, thus changing the climate.

Science has been the basis for environmental policy since the end of the nineteenth century and the first half of the twentieth when scientists such as Svante Arrhenius, William S. Jevons, Vladimir Verdansky and George E. Hutchinson alerted the public to global warming, the depletion of fuels (especially coal), the intrinsic relation between life and the development of the earth, as well as the influence of human activities in changing the environment. After the Second World War, other scientists started to warn of changing weather conditions, the impacts of overpopulation on world resources and the overproduction of waste.[77] In the early 2000s, decision-makers and the general public started to understand the basics of what the policies address. However, as Ann Campbell asserts, very few studies link policy with science, because political science and international relations tend not to pay attention to science and vice versa.[78] However, decision-makers and their constituencies need to understand the logic behind policy choices. For environmental topics, this relation between science and policy is crucial and has been the basis of climate global governance. In general, climate policy (as almost all kinds of policy) agrees to rely on science. Policy-makers have to interact with it and contest it only when contradictory messages reach the negotiation tables and the public realm.

In this sense, decision-makers and social scientists need to understand the basis of climate science in order to be consistent with the natural trend and ecological situation of the territories they address.

How Does the Climate System Work?

The 4.5-billion-year history of the earth has been characterized by constant change: change in its orbit and rotational movement, change in the shape, size and location of the continents, change in the composition of the atmosphere and in oceanic circulation, change in the type of species inhabiting the biosphere and, of course, change in the distribution of climatic patterns. Climate, like weather, is far from static. Just as weather changes from day to day, climate changes too, but in different timescales, from years, decades and centuries to millennia or even longer; this corresponds with the geological history of the earth.

Climate has swung dramatically, from glacial to interglacial episodes of the Pleistocene (most of the past two million years), and from the much earlier

77 To address these topics, see Arrhenius' book of 1896, *On the Influence of Carbonic Acid in the Air upon the Temperature of the Ground*, or William S. Jevons' famous book *The Coal Question*, first published in 1865. See also Verdansky's 1926 book *The Biosphere* and G.E. Hutchinson's *The Ecological Theater and the Evolutionary Play* (1965). See also Campbell, 2009.

78 Campbell, 2009.

periods of the so-called "snowball earth" to the "global sauna" of the mid-Eocene (roughly 50 million years ago). Organisms responded to these changes through either adaptation or extinction.[79] This constant change eventually gave way to the conditions that rule our planet and allowed our species to develop. Most importantly, over the past 10,000 years (an interglacial period called the Holocene) climate has been relatively stable and favorable for human civilizations to flourish and prosper.

Even during this stable period, there have been regional climatic fluctuations produced by natural forcing, such as the "Little Ice Age" (AD 1600–1800), when Europe experienced unusually cold conditions, or between AD 800 and AD 1000, when there was a series of droughts in the Yucatan Peninsula.[80]

Modern climatic records have shown that the average surface air temperature of the earth has increased by approximately 0.6 °C over the past century and by about 0.2 °C to 0.3 °C over the last 40 years. Studies of the statistical significance of the global surface air temperature trend over the last century have detected a significant change and have shown that the observed warming trend is unlikely to be entirely natural in origin.[81]

History has taught us that variations in the earth's climate have had considerable impacts on society. But just as organisms are affected by climatic conditions, organisms can act as "ecosystem engineers," also affecting their environment. Some species do it locally, on a small scale, such as beavers. Others, such as cyanobacteria, can have profound effects in changing the composition of the atmosphere (producing oxygen) over millions of years. Now the human race is altering the composition of the atmosphere, first on a local scale, and since the industrial revolution at the end of the nineteenth century, on a global scale.

How will this new anthropogenic forcing affect climate variability? And, most importantly, what effects will such climatic variation have on the planet and on human development? Predicting the effects that a global increase in temperature could have on our lives is no simple task. Thus we need a better understanding of how the climate system operates globally and regionally.

In order to understand global climate change, we must first define climate and the way it operates. Climate acts as the primary control for ecosystem distribution.[82] The physical conditions that make life possible on earth are determined by our planet's location relative to the sun and the greenhouse effect of its atmosphere.

Climate is conventionally defined as the long-term statistics of the weather (temperature, cloud cover and precipitation). These physical processes within the atmosphere are affected by several factors, such as ocean circulation, the reflectivity of the earth's surface, the chemical composition of the atmosphere and vegetation patterns. However, the driving force that controls these conditions is energy. There are three main sources of energy: (1) solar radiation providing heat

79 Lomolino et al., 2006.

80 Bradley, 2000.

81 Bertrand et al., 2002.

82 Bailey, 1998.

and light; (2) the kinetic energy of the rotation and orbit of the earth; and (3) the internal forces of both heat and kinetic energy together.[83]

The solar radiation the earth receives is continuous. Some of this energy is absorbed at the earth's surface or by the atmosphere, while some is reflected back to space. At the same time, the earth and its atmosphere emit energy to space, resulting in a balance between energy received and energy lost.[84]

The energy arriving at the earth comes in different forms, such as ultra-violet and visible radiation. The earth emits back terrestrial radiation in the form of infra-red energy. These two major energy fluxes must balance, in other words, stay in equilibrium. The earth's atmosphere, however, affects the nature of this energy balance. In normal conditions, the greenhouse gases allow the short-wave solar radiation to pass through unimpeded, while trapping (absorbing) most of the long-wave terrestrial radiation. As a result, the average global temperature of the earth has been 15 °C, 33 °C warmer than it would be without an atmosphere.[85] Without GHGs in the atmosphere, global temperature would be about–17 °C and a blanket of ice would perpetually cover the earth.[86] A latitudinal interchange of heat also exists from the tropics to the poles. If this energy transfer did not occur, the equator would be 14 °C warmer on average, while the North Pole would be 25 °C colder.[87]

The atmosphere does not behave as an isolated system. Transfers of energy take place between the atmosphere and the other parts of the climate system. The most important of these include the world's oceans, where surface ocean currents assist in the latitudinal transfer of heat. Warm water moves towards the poles, while cold water returns towards the equator. Energy is also transferred via moisture. Water evaporating from the surface of the oceans stores heat, which is subsequently released when the vapor condenses to form clouds and rain. The surface ocean currents form part of a wider global ocean circulation system, with deep-water currents also transferring energy across the earth. This system is called the global thermohaline circulation and, as its name suggests, it is driven by differences in temperature and salinity.[88]

The second element of energy flux that drives climate is earth's rotation. As a result of the way the earth revolves about the sun and rotates on its axis, the low latitudes or tropics receive more solar radiation than do middle and higher latitudes. Only about 40 percent of solar energy is received above the poles compared to the equator. To balance this energy, there is a large-scale transfer of heat to the poles, which is accomplished through atmospheric and oceanic circulation. The frictional effects of the rotating earth's surface on air flow cause circulation to be relatively complex. Nevertheless, the solar energy and atmospheric and oceanic

83 Ibid.

84 Jacob, 2005.

85 Buchdahl, 2002; Cammack et al., 2008.

86 Sterman and Sweeney, 2002.

87 Buchdahl, 2002.

88 Ibid.; Cammack et al., 2008.

circulation are distributed over the earth in an organized manner. These controls produce recognizable world patterns of temperature and precipitation, the two most important climatic elements.[89]

The third element is the balance among energy fluxes. Despite the complexity of the climate, the essentials are simple and can be easily understood with basic knowledge of stocks and flows. In this sense, the earth receives a continuous influx of energy from the sun. Some of this energy is absorbed at the earth's surface or by the atmosphere, while some is reflected back to space. At the same time, the earth and its atmosphere emit energy to space, resulting in an approximate balance between energy received and energy lost.[90] The warmer the earth, the greater the flow of energy radiated back into space. Incoming solar energy heats the earth until it is just warm enough for the energy radiated back to space to balance the solar energy input. The amount of energy radiated back into space depends on the composition of the atmosphere.[91]

The Atmosphere and Greenhouse Gases

The atmosphere is a mixture of different gases and aerosols (suspended liquid and solid particles). The two most abundant gases are nitrogen (78 percent by volume) and oxygen (21 percent by volume). However, it is the greenhouse gases (GHGs), despite their relative scarcity, that play a more determinant role in the regulation of the atmosphere's energy budget. These include carbon dioxide (364 parts per million volume—PPMV), methane (1720 parts per billion volume—PPBV) and nitrous oxide. In addition, water vapor makes up about 2 percent of the global atmosphere, and is itself the most important of the natural greenhouse gases.[92] Table 2.2 shows the natural and the synthetic GHGs, as well as the type of products in which they can be found.

As shown in Table 2.2, these gases have different warming potentials. Although CO_2 has the lowest warming potential, it is the most abundant gas. On the other hand, SF_6 is scarce, but one unit would warm 23,900 times what it could warm by CO_2. GHGs are very efficient at absorbing the longer wavelengths of infra-red radiation emitted by the earth and kept in the atmosphere. As mentioned above, the trapping of heat by these gases controls the earth's surface temperature, despite their presence in only small concentrations in the atmosphere. Certain

89 Bailey, 1998.

90 Jacobs, 2005.

91 Sterman and Sweeney, 2002.

92 Water vapor, the most abundant greenhouse gas, is thought to increase in concentration in response to increased concentrations of the other greenhouse gases as a result of feedbacks in the climate system. See Jacob, 2005. However, water vapor was not included in the six GHGs dealt with in the UNFCCC, due to the fact that it is not a pollutant gas and because of the difficulty of addressing responsibility for its emission (which can be natural or human).

Table 2.2 Greenhouse gases, sources and warming potential

	Gases	Sources	Warming Potential
Natural origin	Carbon dioxide (CO_2)	Combustion of fossils (coal, oil, petroleum, gas), chemical reactions in manufacturing (cement and steel), land-use change and forestry	1
	Methane (CH_4)	Anaerobic decomposition (agriculture, waste deposits, animal excrement), gas leaks	21
	Nitrous oxide (N_2O)	Production and use of nitrogen fertilizers, combustion of fossil fuels	310
Anthropogenic Origin	Hydrofluorocarbons (HFCs)	Manufacturing, especially for cooling systems	140–11,700
	Perfluorocarbons (PFCs)	Manufacturing, especially for cooling systems	6,500–9,200
	Sulfur hexafluoride (SF_6)	Manufacturing, especially when a dielectric fluid is needed	23,900

Source: Instituto Nacional de Ecología, 2011.

atmospheric gases absorb radiation at some wavelengths but allow radiation at other wavelengths to pass through. The atmosphere is mostly transparent (with little absorption) in the visible part of the spectrum, but significant absorption of solar ultra-violet radiation by ozone, and terrestrial infra-red radiation by water vapor, carbon dioxide and other trace gases occurs.[93]

The absorption of terrestrial infra-red radiation is particularly important to the energy budget of the earth's atmosphere. Such absorption by GHGs heats the atmosphere, and so the earth stores more energy near its surface than it would if there were no atmosphere. This process is popularly known as the "greenhouse effect." Glass in a greenhouse is transparent to solar radiation, but opaque to terrestrial infra-red radiation. The glass acts like some of the atmospheric gases and absorbs the outgoing energy. Much of this energy is then re-emitted back into the greenhouse, causing the temperature inside to rise, as shown in Figure 2.1.

93 Buchdahl, 2002.

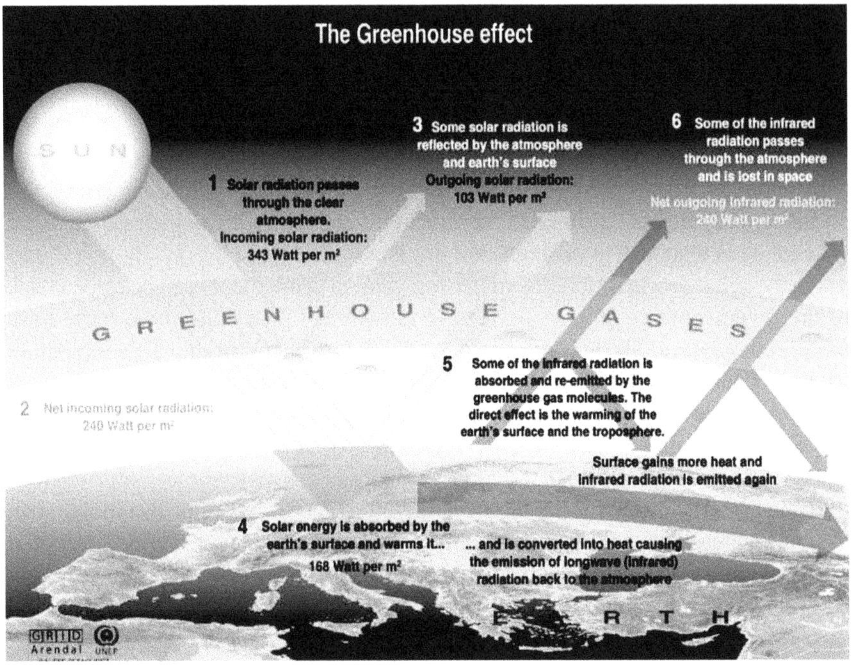

Figure 2.1 The greenhouse effect
Source: Intergovernmental Panel on Climate Change, 2007a.

Thus an increase in GHGs causes the warming up of the surface of the earth (especially oceans). This situation interferes with the natural flux of energy and affects climate patterns.

Climate Change, Variability and the Anthropogenic Factor

Both natural and anthropogenic forcing plays a role in explaining the overall global warming of the last 150 years. This must be kept in mind when trying to project future climate changes resulting from human activities. These changes need to be set in the context of what has happened in the past. On a century-basis scale, changes in solar irradiance have played an important role in the temperature variations of pre-industrial times. They have the potential to account for climates as cold as the Little Ice Age or as warm as the medieval epoch.[94]

However, the relatively short instrumental record of climate clearly does not represent a steady background against which future variations can be gauged. Human-induced change will be difficult to assess unless the long-term natural variability of the climate system can be characterized. Natural variations

94 Bertrand et al., 2002.

with timescales of decades to centuries may well be masking anthropogenic climate changes that are already in place—and will remain. We must be able to recognize natural variability and its results if we are to make reasoned estimates as to whether a particular climate perturbation or trend is likely to have been induced by human activities, or simply represents a natural variation.[95] Whatever anthropogenic effects there are on climate in the future, they will be superimposed on the underlying background of "natural" climate variability, which varies on all timescales in response to different forcing factors.[96]

Although a few instrumental meteorological measurements can be traced back to the seventeenth century, no extensive network of data exists before the mid-nineteenth century. The record of large-scale (hemispheric or global) temperature extends back only 150 years. Even the longest instrumental records cover barely 300 years.[97] Hence, although we may have a 300-year record of climate at a few specific locations, we have only approximately a 150-year perspective on the spatial–temporal variability of the earth's climate system as a whole. Unfortunately, the time for which we have detailed instrumental records happens to coincide with the time when human activities have increasingly polluted the atmosphere on a global scale. If we want to understand the relationship between any underlying natural forcing and climate variability, we must probe further back in time, and for this we must rely on paleoclimate archives (called climate proxies).[98]

The identification of the characteristics of climate variability involves several issues. Differentiating climate "change" from climate "variability" is a matter of timescale. What appears to be a trend, in a single decade's recording, may reveal itself as fluctuating variability over a period of a century. The Dust Bowl period in the central USA in the 1930s represents a short, decadal-scale natural variation in climate for a specific region. In contrast, the Little Ice Age, which lasted from the 1400s to the 1800s, represents a variation on a timescale of centuries.

An additional difficulty in assessing variability is that change and variation are often characterized by strong spatial dependency. That is, while some region or regions of the earth may be experiencing significant climate variation over a given period, other regions may show virtually no change. Similarly, a specific variable (e.g. global temperature) may show a marked change between decades, while another (e.g. global pressure fields) may not show any significant difference. Where and how climate is measured can influence the findings and conclusions.[99]

Natural biogeochemical processes have caused the concentration of CO_2 in the atmosphere to fluctuate over geological time, and surface temperatures have fluctuated with it. General measurements of the Intergovernmental Panel on Climate Change assert that human activity has now reached a scale where it

95 National Research Council, 1995.

96 Bradley, 2000.

97 Ibid.

98 Ibid.

99 Jacob, 2005.

affects climate processes significantly. Anthropogenic GHG emissions have been growing exponentially since the beginning of the industrial age. Some factors give a basis to this growth in GHGs.

The industrial revolution in the nineteenth century has allowed human population to grow from one billion worldwide in 1800, to two billion in 1930 and to more than seven billion in 2013.[100] This has created the need for more production of goods and services, with the consequent exploitation of natural resources and raw materials. Toxic substances are being poured into water streams; forests have been decimated all over the world and, with them, the services they provide; the extensive use of fossil fuels and synthetic gases has produced an increase in the atmospheric concentration of GHGs.

Concentrations of CO_2 and other GHGs including nitrous oxide (N_2O), methane (CH_4), chlorofluorocarbons (CFCs), hydrofluorocarbons (HFCs), perfluorocarbons (PFCs), and others have been growing exponentially. For 420,000 years of history, the pre-industrial atmospheric CO_2 concentration was about 280 PPM; today it is almost 450 PPM and rising.[101]

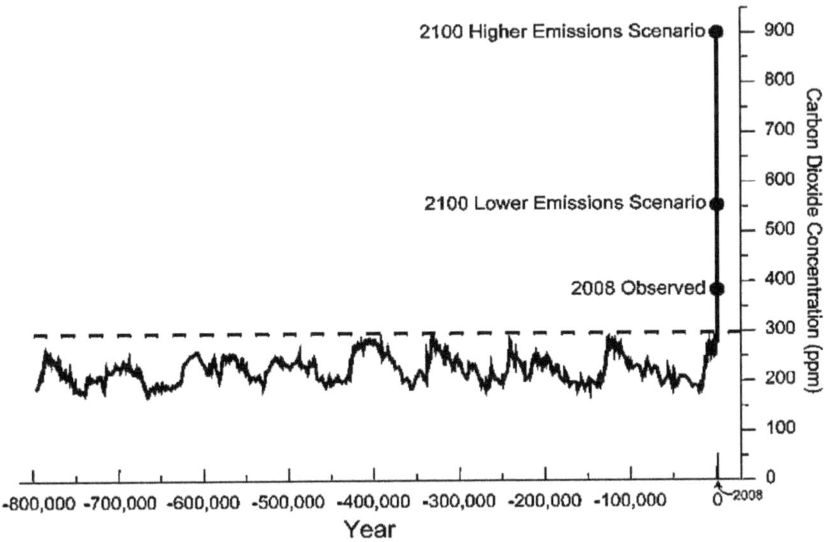

Figure 2.2 Concentration of CO_2 over time

Source: National Climate Data Center, 2013.

100 United States Census Bureau, 2013; United Nations Population Division, 2011.

101 International Energy Agency, 2010.

Concentrations of other important GHGs are also rising: for example, NO_2 and CH_4 concentrations are up by 17 percent and 151 percent, respectively, since 1750. Current GHG concentrations contribute about 2.4 watts per square meter of net radiative forcing; that is, incoming solar radiation exceeds outgoing radiation by 2.4. Consequently, global surface temperature has been rising.[102]

Several scholars have concluded that the rise in temperature during the twentieth century (particularly the late twentieth century) is unprecedented, and the spatial pattern of warming cannot be explained by natural forcings alone; increases in GHGs (natural and synthetic) provide the necessary additional forcing to account for the recent changes in temperature.[103] Climate simulations have indeed shown that human activities (those that modify the atmospheric concentration of GHGs) increase the models' ability to simulate the spatial and temporal pattern of the historical temperature peaks.[104] Today there is no doubt that the composition of the atmosphere and the reactions between the atmospheric chemical constituents are changing as a result of human activity.

For example, we have been burning organic carbon stored in the biosphere as coal and oil, which was deposited in past geological times. As a consequence, the atmospheric concentration of CO_2 has been rising, basically because the rate of combustion of fossil carbon is higher than the rate of absorption by other compartments of the earth's biogeochemical system.[105]

Placing the short instrumental record of climate into a longer-term model, it is evident that numerous paleotemperature records reveal that the twentieth century has been exceptionally warm in the context of the last millennium, and perhaps many millennia. Furthermore, the coldest decades of the last century were among the coldest times in the late Holocene period (which is the contemporary one). Thus the world has experienced both the warmest and the coldest extremes of the late Holocene within a brief interval of fewer than 200 years.[106]

When overlapping the rate of CO_2 emissions with the temperature, the data coincide and present increases since the twentieth century.[107] This overlapping has been the main argument for establishing climate global governance and institutions such as the United Nations Framework Convention on Climate Change in 1992. Chapter 3 provides a detailed discussion on the role of the scientific community regarding climate governance and the legitimacy of institutions, such as the Intergovernmental Panel on Climate Change. Facing the evidence presented by the scientific community, a decision-maker has the political choice of not prioritizing the topic or creating mechanisms of climate governance. This

102 Sterman and Sweeney, 2002; Intergovernmental Panel on Climate Change, 2007a.

103 Battarbee and Binney, 2008; Crowley, 2000; Intergovernmental Panel on Climate Change, 2007a.

104 Bertrand, 2002.

105 Schlesinger, 2000.

106 Bradley, 2000.

107 Environmental Protection Agency—US, 2011.

section briefly showed how GHGs behave and why the variation in the density of these gases impacts the climate system. Chapter 3 resumes this discussion of how the epistemic community (not only scientists, but also social scientists) has influenced the agenda setting for the provision of global public goods under climate governance institutions.

Chapter 3

Limits to Global Climate Governance: The Failure of the Kyoto Protocol

Introduction

Environmental governance started to develop in the 1970s. At that time, international responses reacted to pollution problems, especially in the US, Canada and Europe. The first approach was regulatory, and it focused on single-pollutant issues that were thought to be solved by technology.[1] Later, diverse responses emerged. In the following decades, definitions of environmental governance went from the need to commoditize environment and develop international rules and law, to the need to reform the previous command-and-control regulatory frameworks and democratize them on a communitarian basis. In addition to these approaches the "third way" for conceiving environmental governance emerged. This focused on the need for accountability of environmental results, rather than accountability for compliance with rules and regulations.[2]

In this context, several environmental issues were prioritized. One of them was air pollution and its impact on climate. It is the goal of this chapter to provide a background to global climate governance and explore specific mechanisms, such as the United Nations Framework Convention on Climate Change (UNFCCC) and the Kyoto Protocol (KP). The first section presents the background to developing global strategies to deal with climate change. A historical analysis is presented on how the topic reached the international agenda. The second section explains how the KP was designed and outline all its structural elements. The third section discusses why this mechanism has not worked as expected. The general argument of the KP failure is threefold. First, the KP rests on the idealistic assumption that national policies would have to adapt to global needs. Second, the Protocol is a mechanism with an inadequate design based on a national approach, when in reality the implementation costs are localized regionally. The second explanation deals with the emergence of free-riders. This argument was developed by the US Congress for not ratifying the KP. The explanation was simple. Some potential polluting countries (which might be growing fast but leaving a huge GHG footprint), such as China and India, were not included in the "Annex I" list. Thus the US and the other Annex I countries would absorb the greatest economic environmental

1 Durant et al., 2004:1.

2 Ibid.:2.

costs of these two countries. Third, the KP has no obligatory mechanisms for implementation; it is based on voluntary instruments, targets and commitments.

Global Climate Governance: Towards Kyoto and Beyond

With 195 Parties, the UNFCCC has near full membership to date (if compared with the members of the UN system). This treaty is the framework for the 1997 KP, which up to the end of 2012 counted 193 member Parties (192 countries and one regional organization, the European Union). Under the Protocol, 37 states were tasked with specific goals to reduce or limit their GHGs.[3] However, from December 2012, one of the Parties (Canada) withdrew from the Protocol.[4]

The journey that finally resulted in this agreement started after the Second World War in 1957 with the celebration of the International Geophysical Year, when scientific studies of global warming were first drafted.[5] More than a decade later, at the 1972 UN Conference on the Human Environment held in Stockholm, a set of principles was established to protect environmental depletion produced by human activities. This was the first international UN conference at which environmental topics were placed on the negotiations table. This conference established that developed nations would bear greater responsibility regarding environmental and inequality issues in other countries. After this conference, the United Nations Environment Program (UNEP) became the voice for the environment in the UN system.[6] The UNEP, along with the World Meteorological Organization (WMO) and the International Council of Scientific Unions (ICSU), organized the World Climate Conference in Geneva in 1979.[7] At this first international conference attention was drawn to topics such as acid rain, toxic waste and the ozone layer, especially in urban industrialized areas. Greenhouse gases, though recognized as pollutants, were omitted from the agenda.

In the 1980s, climate change moved from the scientific agenda to the policy agenda, both nationally and internationally.[8] It was with the Vienna Convention for the Protection of the Ozone Layer in 1985 and its Montreal Protocol on Substances that Deplete the Ozone Layer in 1987 that climate change started to take its own path and became a global issue. The scientific community had a great influence in placing the topic on the international agenda. In 1986, two reports alerting the public to global warming and climate change were presented. The first one was designed by the ICSU, the UNEP and the WMO; the second one by the National Aeronautics and Space Administration (NASA). Both reports talked about the

---

3 United Nations Framework Convention on Climate Change, 2011d.

4 United Nations Framework Convention on Climate Change, 2012.

5 Adams, 2009:99. See also the section on the science of climate change in Chapter 2.

6 United Nations Framework Convention on Climate Change, 2007b.

7 Adams, 2009:99.

8 Weaver, 2004.

uncertainty of the increase in GHGs to the atmosphere. These documents set the basis for drafting technicalities in international agreements.[9] For example, in the drafting of the Montreal Protocol, scenarios for future emissions of all GHGs were discussed in depth,[10] especially their impact on the ozone layer, which already presented evidence of depletion: an expanding hole over the North Pole was found. The 1987 Montreal Protocol intended to control chlorofluorocarbons (CFCs) and hydrochlorofluorocarbons (HCFCs), synthetic compounds that depleted the ozone layer and were GHGs. The result of the 1987 ozone summit was that CFCs and HCFCs were replaced worldwide by hydrofluorocarbons (HFCs) and perfluorocarbons (PFCs); these gases had lower warming potential and would cause less damage to the ozone layer.[11]

Ozone layer depletion was the "hot topic" of that decade and "signaled that environmental issues were increasingly moving from local and national to the global level."[12] The Vienna Convention and the Montreal Protocol served as a model for the next generation of global governance conferences on climate change, represented by the 1988 conference called "The Changing Atmosphere: Implications for Global Security." It was held in Toronto, just after the Intergovernmental Panel on Climate Change (IPCC) was created.[13] The IPCC had the goal of legitimizing global climate governance by presenting assessments from three working groups that addressed the science, the socio-economic and vulnerability-adaptation issues, and the socio-economic and mitigation strategies. In addition, the IPCC mandate, received from the UN, established that the organism had to set the basis for understanding the risk of human-induced pollution and its impacts, as well as options for mitigation and adaptation.[14] However, the IPCC did not have a research team of its own; it worked only as a peer reviewer of literature of worldwide experts.

The Toronto Conference set climate change into the international debate with the legitimacy granted by the newly created epistemic community. Legitimacy, then, would fall under "networks of professionals with recognized expertise and competence in a particular domain and an authoritative claim to policy-relevant knowledge within that domain."[15] This fact attracted important personalities such as Gro Harlem Brundtland, Chair of the World Commission on Environment and Development, and over 300 scientists and policy-makers from 46 countries

---

9 For a detailed analysis of the agenda-setting capacities and the scientific narratives regarding climate change from a historical perspective, see Campbell, 2009.

10 Adams, 2009:99.

11 CFCs and HCFCs were used as cooling devices in the refrigeration industry, vehicles, appliances, as well as in aerosols. See United Nations Framework Convention on Climate Change, 2009b.

12 Paehlke, 2008:58.

13 The IPCC was created in 1988 by the WMO and the UNEP. See Bramley, 2000:1–2.

14 Weaver, 2004.

15 Haas, 1992.

and organizations. This conference called for reductions of carbon dioxide (CO_2) emissions by 20 percent from 1988 levels by the year 2005.

The Vienna and the Toronto conferences complemented each other. The cooperation model of Toronto was based on the Montreal Protocol; it was expected to be as successful for climate change as it had been for the ozone layer. The Toronto Conference included HFCs and PFCs in the list of anthropogenic-designed GHGs and set a major precedent for widening the scope of the Montreal Protocol. As established by Robert Paehlke, the Toronto Conference coincided with

> the Second Wave of Environmentalism (roughly 1986–91), a period when public opinion and media attention focused unrelentingly on environmental matters—having being spurred by the issues of acid precipitation, the 1987 media sensation concerning the global journey of a barge filled with garbage from New York City, the 1989 Exxon Valdez oil spill, [depletion] of tropical rainforests, and growing scientific evidence about human-induced climate change.[16]

After the Toronto Conference, representatives of countries such as Canada, the US, and some of the most developed countries in Europe, started talking about setting specific goals for stabilizing CO_2 emissions as a basis for the UN meeting in Bergen, Norway, to be held in 1990.[17] At this point, the climate-change conference had been a success in gathering international representatives and putting the topic on the international agenda. It was applauded by several actors and encouraged the participation of society in this kind of topic. For example, Canadian governmental organizations, green NGOs, universities, think tanks and industries participated through Canada's National Climate Change Process, which included a long series of public hearings to define the country's position at the meeting.[18]

At the beginning of the 1990s, some countries expressed the need for an international climate-change agreement. They talked about the necessity of moving towards the "next generation" of environmental policies that would address both point sources (for example, chimneys) and nonpoint sources (such as pesticides) and adopt a source-reduction approach.[19] In other words, the pledge was to design mitigation strategies. In the same year, the IPCC presented its first report and established the importance of cutting fossil fuel and carbon consumption to address air pollution. This recommendation created conflict between industrialized nations and developing ones. The North was "urging the priority of environmental protection and that any measures agreed should be cost effective, while the South pushed the need for development and industrialization, and the principle of historical responsibility."[20]

---

16 Paehlke, 2008:59.
17 Smith, 1998:4–5.
18 Paehlke, 2008:64.
19 Van Nijnatten, 2008a:294.
20 Adams, 2009:102.

The international context made these years conflictive and confusing. With the approval of the UN General Assembly, the Second World Climate Conference took place in Geneva in 1990 with the objective of creating an agreement to present at the coming Rio Conference in 1992. However, there were very different positions towards GHG emissions. "The EU favoured an agreement on targets and a timetable for implementation, the USA was reluctant."[21] In the end, the compromise to be presented in the context of the Rio Conference was to set a non-binding agreement for cutting CO_2 and other GHGs not treated by the Montreal Protocol to 1990 levels (the year in which negotiations had started).

The Rio Conference was held in 1992. The world was expecting great results. At this meeting the previous agreement was introduced as the United Nations Framework Convention on Climate Change. Needless to say, this was only one of several international agreements of equal importance, such as Agenda 21, the Rio Declaration on Environment and Development or the Convention on Biological Diversity. Although the majority of the conventions proposed in the context of the Rio Conference had the same logic (frequent meetings to follow up implementation of the Convention, financial instruments and implementation mechanisms), the Convention on Climate Change had its own dynamic. The institutional architecture to be designed was sustained, and worked based on the assertion of the previous IPCC reports that demonstrated that climate change was in part the result of human activities. Their general goal was to establish mechanisms to incorporate the climate issue into a sustainable development approach, a concept developed since the 1987 Conference[22] and reinforced at the Rio Conference in 1992. The concept of sustainable development also included the need to preserve the environment for future generations. The UNFCCC established the main goal to achieve the

> stabilization of greenhouse gas concentrations at a level that would prevent dangerous anthropogenic (human induced) interference with the climate system. Such a level should be achieved within a time-frame sufficient to allow ecosystems to adapt naturally to climate change, to ensure that food production is not threatened, and to enable economic development to proceed in a sustainable manner.[23]

The UNFCCC also decided which gases were to comprise the mix of greenhouse gases—including the Montreal Protocol's GHGs—and the ways to measure them,

21 Ibid.

22 The 1987 Conference was called "the Brundtland Commission" in honor of Gro Harlem Brudtland, Prime Minister of Norway, who promoted it. The official name is the World Commission on Environment and Development. The result of this meeting was a document called *Our Common Future*, in which the concept of sustainable development was defined. See Adams, 2009.

23 Article 2 of the United Nations Framework Convention on Climate Change, 1992:4.

setting 1990 as the baseline year.[24] The selection of the six GHGs was based on the IPCC's scientific reports previously drafted.

Once the need for international regulation was established, principles had to be agreed. The main principle of this treaty was a "common-but-differentiated" approach for responsibility and capabilities.[25] This meant that industrialized countries (OECD members) were to have greater responsibility for controlling GHGs because they had historically developed their economies by polluting (for example, burning fossil fuels for industry). In other words, "they are the source of most past and current greenhouse gas emissions."[26] The rest of the countries ratifying the Convention agreed to take climate change into account in such matters as agriculture, industry, energy and natural resources. In other words, they agreed to develop national programs to slow down climate change through stabilizing GHGs to non-risky levels. The document also recognized that all efforts would be based on national sovereignty.

This convention also made the basic differentiation between Annex I countries and Non-Annex I countries—a distinction that would be enforced in the Convention's protocol in 1997. Countries with specific commitments and time goals, either to reduce specific GHG percentages or to limit them, would be included in the list called Annex I. This Annex also included countries with transition economies, especially those in Eastern Europe. Annex Parties would have major responsibilities regarding financial mechanisms for climate change. The rest of the signing countries did not appear in the annexes. These non-annex countries were developing countries with no reduction or limit commitments. This differentiation was also present when the UNFCCC pointed out that the extent of climate change was uncertain because it was not known what impact GHG releases would have. Developing countries would have relatively greater climate-change impacts, although their emissions were the lowest. Due to this situation, they were not committed but encouraged to propose projects on a voluntary basis, especially in the area of adaptation. The mechanism for financing the climate-change strategies was also established at the UNFCCC. An important part of this climate-change agreement was Article 14, which provided mechanisms for dispute settlement to address misunderstandings, lack of accountability or transparency, and weak implementation capacity. This article established that there were two institutions for solving controversies: the International Court of Justice and a

24 The UNFCCC established this year as a baseline, but in Article 3(5–8) of the KP, countries may propose a different base year for measuring GHG reductions. However, if a country chooses some posterior base year, the GHGs produced would have to be multiplied by five. Any Party included in Annex I may use 1995 as its base year for HFCs, PFCs and SF_6. See the KP at United Nations Framework Convention on Climate Change, 1998:4.

25 Article 3–4 of the United Nations Framework Convention on Climate Change, 1992:4–5.

26 United Nations Framework Convention on Climate Change, 1992:1.

process of arbitration by the Conference of Parties (COPs) mechanism. The COPs were the multilateral mechanisms whereby polluters and non-polluters would sort out their differences and find ways to compensate each other.

COPs Become Important: From Kyoto (1997) to Doha (2012)

Under Article 17 of the UNFCCC, the adoption of protocols is allowed under COP meetings.[27] The Convention entered into force in March of 1994; the meetings started in 1995. The first reunion took place in Berlin (COP-1), with the goal of developing quantified limits and targets, and specific time frames.[28] At this meeting, some key institutional components were established, such as Subsidiary Bodies for Scientific and Technical Advice and for Implementation. COP-1 also confirmed the Global Environmental Facility (GEF) as the financing agency (administered by the World Bank at that time). This meeting started to clarify the positions of different countries regarding specific reduction commitments set for the year 2000. As Bartsch and Müller note, from the beginning these goals were not realistic. On the one side were the OPEC countries and US industry groups claiming the impossibility of achieving the targets. On the other was the Alliance of Small Island States (AOSIS) calling for 20 percent reduction by industrialized countries by 2005.[29] Facing this gridlock, the COP members enacted the Berlin Mandate. This document mandated the creation of an ad hoc group to negotiate and draft a scheme that would satisfy all members, taking no longer than two years.

At the following meeting, COP-2 in Geneva in 1996, the US delegation introduced the idea of pricing pollution through trading schemes. The US was accountable for more that 20 percent of GHG emissions at that time, and, as mentioned by Meckling and Hepburn, it preferred a market-based policy. The US had experimented successfully with a domestic sulfur dioxide (SO_2) trading scheme, which had been highly cost-effective. Lobbies of firms and companies supported the idea. Unexpectedly, an alliance of green organizations also pushed for the trading scheme. For example, there was a strong alliance between British Petroleum and Environmental Defence in favor of the market approach. They saw it as the only way to pressure firms into discussion and implementation of GHG emission targets. This new advocacy strategy gave a basis for emissions trading under Clinton's government, and it would be a cornerstone of the flexible mechanisms of the forthcoming KP.[30]

In December of 1997, 160 members of the UNFCCC met in Kyoto, Japan, for the COP-3 to develop an international accord that would help mitigate the effects of climate change through the reduction of GHG emissions. In the following years,

27 United Nations Framework Convention on Climate Change, 1992:19.

28 A summary of the main achievements of the COPs to date is presented in Appendix B of this book.

29 Bartsch and Müller, 2000.

30 Meckling and Hepburn, 2013:476.

the resulting agreement, called the Kyoto Protocol, was signed by the majority of the countries; it was subject to individual ratification, entering into force

> on the ninetieth day after the date on which not less than 55 Parties to the Convention, incorporating Parties included in Annex I which accounted in total for at least 55 per cent of the total carbon dioxide emissions for 1990 of the Parties included in Annex I, have deposited their instruments of ratification, acceptance, approval or accession.[31]

Thus the KP became a legal instrument subsidiary to the UNFCCC. Under the terms of the Protocol and based on the listing of countries already established at the UNFCCC, Annex I countries committed under Articles 2 and 3 to implement and elaborate policies and measures in accordance with their national circumstances to reduce anthropogenic emissions of at least 5 percent below levels of 1990 for 2008–12.[32] As an interim step towards reporting under the KP, Annex I countries were required to submit annually a National Inventory Report (NIR) and a set of data from 1990 onwards to the UNFCCC Secretariat on 15 April of each year. These submissions had to follow a set of internationally agreed guidelines developed by the IPCC for the preparation of national GHG inventories.[33] Article 2 specified that Annex I countries had to do whatever was necessary to use renewable energies and perform technological change towards emissions reduction of GHGs not treated in the Montreal Protocol of 1987.

Right after the establishment of the UNFCCC in 1992, negotiations started by setting the positions of countries regarding GHG reductions. The general perception was that the US would establish a 3 percent GHG reduction. Most of the Annex I countries, especially Canada, the European Union and Japan, started making commitments around the US percentage. Facing this scenario, two clear strategies within Annex I countries appeared.

The first one was the regional approach taken by the European Union. The KP allowed this regional approach, also called "bubbling." This mechanism permitted Parties to redistribute their commitments to facilitate the achievement of their collective target. However, Article 4(5) noted that, "[i]n the event of failure by the Parties to such an agreement to achieve their total combined level of emission reductions, each Party to that agreement shall be responsible for its own level of

---

31 See Article 25 of the KP at United Nations Framework Convention on Climate Change, 1998.

32 The UNFCCC listed countries subject to commitments at Annex I. This list was transferred to the KP into Annex B and specific percentages were added to those commitments. To avoid confusion, in this book the committed countries are called "Annex I" countries. See the KP at United Nations Framework Convention on Climate Change 1998:3–4. See also Appendix A of this book.

33 See Article 5 of the KP at United Nations Framework Convention on Climate Change, 1998:6.

emissions set out in the agreement."[34] Nonetheless, Europe chose to negotiate in this way to compensate polluting countries of the Mediterranean (such as Spain, Portugal, Italy and Greece).[35]

The second strategy was to adopt individual commitments. At the beginning, countries such as the US set a 3 percent commitment. All other countries established their own commitments very close to the American one. This approach was taken by Canada, where "negotiators led the delegation to stay 1% behind the US. Canada was cognizant of the need to remain economically competitive with its largest trading partner."[36] Canada would set a goal of 2 percent.[37] This country had had a free trade agreement with the US since 1989 and had entered the North American trading project in 1994 (including Mexico). This represented an integrated economy in production and services, and explained the link to the US Kyoto goals.

In this context of perceptions and strategies, Al Gore, vice president of the US, "suddenly announced that the US, the world's larger emitter, would reduce its emissions by an astonishing 7% from 1990 levels."[38] As a result, many countries had to raise their targets. In addition, during the KP international negotiations, a group led by the Clinton administration favored flexible mechanisms such as carbon trading, sinks and exchangeable credits for developing countries (the US, Norway, Australia, New Zealand, Iceland, Japan and Canada).[39]

However, electoral timing interfered with the US leadership of Kyoto: George W. Bush won the presidential elections. In spite of the Clinton–Gore demands to keep Kyoto alive, in the spring of 2001 the newly elected president announced that the US would not continue with the Protocol. The US Congress had decided not to ratify it. The strongest reason for this decision was the presence of free-riders. The US Congress explained that potential polluting countries, which were expected to grow fast but leave a huge GHG footprint, such as China and India, were not included in "Annex I."

In this scenario, the US and the other Annex I countries would need to underwrite these countries' responsibilities. This resulted in unwillingness to share the costs of remediating global public bads and spillovers of other countries. At both global and local levels individuals or organizations (such as states) could make credible commitments, but they were frequently finding new constraints that changed the basic structure of incentives.[40] The KP process did not take into account that the paymaster that was going to absorb the highest costs (the US) would suddenly decide not to ratify the agreement; the other Annex I countries

34 See the KP at United Nations Framework Convention on Climate Change, 1998:5.

35 Antal, 2004.

36 Harrison, 2007:103.

37 Jaccard et al., 2007:35.

38 Ibid.:39.

39 Jaccard et al., 2007:64–5.

40 Keohane and Ostrom, 1995:2.

simply were not willing to pay up. In this way, legitimacy was lost because the US broke the consensus.

Faced with this situation, many governments and the business community outside the US wanted to call off the KP and adjust their commitments. Some of the Annex I countries, such as Canada, had to stick to the Protocol because of internal environmental pressures. Some of these countries kept fighting for exceptions and privileges in the following COP meetings.[41] For example, Canada and Japan fought to include carbon sinks for GHG reduction and transferable credits for developing countries. At the COP-7 in Marrakech, Morocco, Canada renewed the old argument for including "clean energy" to count as GHG reduction. The Canadian government pledged to receive credits to export natural gas to the US, on the grounds that Americans would otherwise be burning more GHG-intensive oil or coal.[42]

In 2002, at the Johannesburg Conference on Environment and Development (best known as "Rio plus 10"), the majority of Annex I countries, despite having ratified the KP, had not fulfilled their commitments. From the beginning, Canada and the commercial partners of the US were destined to fail to reach their KP targets. The EU had better scenarios for achieving their committed goals because of the regional approach.[43]

During 2006, the Asia-Pacific Partnership on Clean Development and Climate was formed. This initiative implied that there were other forums in which to establish goals, other than the KP and the UN. The guests at the meeting were representatives from the US, China, India, Australia, South Korea and Japan, and the purpose was to help big polluters such as China and India to continue to industrialize using environmental technologies. Yet this initiative did not set emission targets or deadlines; participation was voluntary.[44]

At the beginning of 2007, the IPCC launched a series of three reports that resulted from the work of scientists all over the world: Working Group I on "The Physical Science Basis," Working Group II on "Impacts, Adaptation and Vulnerability" and Working Group III on "Mitigation of Climate Change." These represented international scientific support for the argument that there were both human and natural drivers of climate change.[45] Although this was a big step towards international consensus, some countries contradicted the report's assertion that some global warming was, indeed, anthropogenic. The case of a group of scientists in the US and Australia was a clear example.[46]

41 Jaccard et al., 2007:67.

42 Harrison, 2007:106.

43 Schwanen, 2006:296; Antal, 2004:157–8.

44 Sheppard, 2006.

45 Intergovernmental Panel on Climate Change, 2007a,b,c.

46 For example Lance Endersbee, Australian researcher, who proposes that climate change is a totally natural phenomenon and humans have no effect on it. See Endersbee, 2008. For further details, see the discussion by McCright and Dulap, 2003. They note that the

At the Bali Conference of that same year (COP-13), the majority of Annex I countries confirmed reluctance to comply with the Kyoto commitments. They tried to block international negotiations for setting new targets for industrialized countries for a post-Kyoto agreement.[47] The only achievement of COP-13 was to draft a "roadmap" for future agreements.[48]

In this context, electoral processes in the US again influenced the trend of climate-change governance. Barack Obama won the elections in the US, and he started his period by facing the financial crisis: trying to set climate change as a government priority was very difficult at that time. In a parallel effort, the Poznan Conference (COP-14) in December 2008 took place with no important advances. Nonetheless, in the aftermath of this conference, some Annex I countries started to set pledges to meet their reduction goals. Just after this conference took place, President Obama visited Canadian Prime Minister Stephen Harper and proposed to create a cap-and-trade system in North America.[49] For President Obama, this would be an important mechanism for technological transition and less dependence on fossil fuels. This position was enforced by the US economic recovery bill signed in February 2010, in which Obama's government granted 76.53 billion USD for renewable energy, energy efficiency and the development of green technology. The Canadian government was also making investments, but only 1.60 billion CAD.[50] Prime Minister Harper was forced to follow the green trend of his southern neighbor in order to try to conform to both countries' environmental rules and standards, thus maintaining their economic relationship.

In 2009, COP-15 took place in Copenhagen, Denmark. Although the conference gathered several chiefs of state and government, the results were not as expected. The final accord pledged to maintain a temperature rise of under 2 °C. It was intended to produce a second version of the KP, but no consensus was achieved. For example, some Non-Annex states that registered increasing GHG emissions, such as Brazil, Russia, India and China (the BRICs), plus Mexico and South Africa, took different postures. China and India denied the possibility of committing to specific reduction goals, even though both had considerably increased their GHGs. Brazil would work to reduce deforestation in the Amazon River Basin by up to 80 percent to 2020; Mexico proposed the Green Fund to help emerging economies to mitigate and adapt.[51] In North America, Canada's posture would clearly depend on the path that the US followed, proposing no specific commitments. Hence several climate

---

conservative movement in the US has made this couter-argument since the 1990s. Tony Clarke, 2008, proposes a similar argument for the case of the US and Canada. He argues that the former lobbyists of the tobacco industry, facing the failure of their cause, changed their focus to de-legitimizing climate change.

47 Demerse, 2008.

48 Giddens, 2010:221.

49 The Economist, 2009.

50 Weiss and Bramley, 2009:1–2.

51 United Nations Framework Convention on Climate Change, 2009a.

bills were debated in the Senate. However, due to the financial crisis, the US mildly reduced GHGs emissions, and some others, such as China, increased their GHGs, positioning the country as number one emitter worldwide, as shown in Figure 3.1.

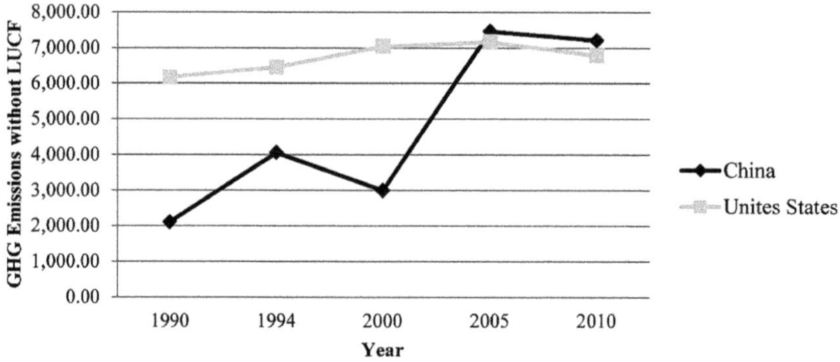

Figure 3.1 US–China greenhouse gases 1990–2010 (MT)
Source: Adapted from United Nations Framework Convention on Climate Change, 2013a.

Albeit modest, the results of the Copenhagen meeting were important in terms of institutional innovation. Apart from committing not to raise the temperature above 2 °C, the developed countries agreed to set scenarios for 2020 and to establish new goals. They also committed to support projects to avoid deforestation through the Reduction of Emissions from Deforestation and Degradation (REDD) mechanisms, and to develop financial instruments such as the Green Fund.[52]

Against this background, COP-16 took place in Cancun, Mexico, facing the challenge of dealing with all the topics not covered at the former meeting. Some examples of these challenges were how to operate the Green Fund, a renewed set of commitments for Annex I countries and the incorporation of some Non-Annex ones, reducing emissions using REDD mechanisms, especially to achieve mitigation. In other words, the goal was to draft a "new" version of the KP.

The Cancun meeting was not as successful as expected either. However, some of the prior goals were discussed and drafted in a package of agreements called "the Cancun Agreements." Among these commitments, COP-16 established that developing countries would have to report emissions every two years. To aid in the technological transfer and funding of these countries, the World Bank would assist in the operation of the Green Fund. The institution would mobilize "fast-start" funding for 2010–12 (30 billion USD) and longer-term funding of 100 billion USD per year to 2020.[53]

52 Ibid.

53 United Nations Framework Convention on Climate Change, 2010b.

The Cancun Agreements also made carbon capture and storage projects eligible for carbon credits under the Clean Development Mechanism.[54] In addition, the REDD mechanisms were included as an alternative mitigation mechanism, as well as adaptation schemes for the most affected countries. The Cancun meeting was based on a renewed scientific assessment by the IPCC, which submitted its conclusions to the US National Academy of Sciences.[55] For 2010, the International Energy Agency noted that the total world emissions of GHGs had increased, mainly from burning fuel. The rise registered was 1.6 Gt (gigatonnes) of CO_2, which represented almost a 10 percent increase over 2009.[56] These projections were made up to 2020, providing evidence that global climate governance had failed.

COP-17 took place in Durban, South Africa, with mixed results. Some of the events taking place at this meeting were very discouraging for global climate governance institutions. Scientific assessments reported that the earth would warm up by 3.5 °C to 2100, surpassing the 2 °C agreed in previous meetings.[57] In addition, all delegations were eager to see China and India commit, but both refused, using the argument of the historical right to pollute. During the Durban Conference, Canada announced its official withdrawal from the KP. The most important reason given was that, after the economic crisis, the country had no chance of achieving the Kyoto goals. In addition, it would not want to continue if the main partner, the US (and now China and India) would not commit. The Canadian withdrawal destroyed the legitimacy of the Protocol and the hopes for COP-17 vanished.

As almost no country had yet accomplished its reduction goals, at the last Conference of Parties (COP-18) in Doha, Qatar, in 2012, the Protocol was granted an extension period of five years in order to develop a document more suitable for the majority of the partners. However, evidence shows no real progress in emissions reduction. A report drafted by the UNEP established that current global GHG emissions to 2012 were estimated at 50.1 $GtCO_2e$, 14 percent higher than the median estimate for 2020 (with a 2 °C target). This is 20 percent higher than emissions in 2000. The report continues by noting that global GHGs are increasing in the context of economic recovery, after a decline during the economic downturn between 2008 and 2009.[58]

Structure of the Kyoto Protocol: Elements of Failure

As mentioned earlier, the KP has been the main instrument of the UNFCCC. It established that the heaviest burden for cutting GHGs fell on the industrialized nations and the emerging economies listed in Annex I. The KP created institutions

54 Friedman, 2010.
55 Busby, 2010.
56 International Energy Agency, 2011.
57 See reports at ibid.; see also The Climate Action Tracker, 2011.
58 United Nations Environmental Program, 2012b.

that relied on several principles. The first is the precautionary principle, which claims that it is better to prevent emitters from producing pollution than to find solutions afterwards. The second refers to common but differentiated responsibilities, based on the premise that the one who pollutes (or has polluted) must pay for the damage (the "polluter-pays" principle). The third deals with sustainable development in the terms defined by the Brundtland Conference of 1984, in which economies may develop, but with a rational use of resources so that future generations are not compromised.[59] The KP legitimized these ideas based on the epistemic communities already created since the IPCC was established.

The need for designing and supporting decisions based on science was fundamental, as Article 13 of the KP mentioned that there was a need to call on the help of private or public entities and members or observer states to create a body of knowledge and information structures.[60] These strategies determined what principles were acceptable to solve disputes and manage conflict, and whether governmental actions were legitimate or illegitimate.[61]

In this context, the lack of legitimacy of the KP (or the constant questioning of its bases) resulted in failure. Legitimacy of global institutions, understood as "the justification of authority,"[62] could be easily lost. The fact that their authority is self-imposed implies that consensus over the contract (or convention), internally and sometimes internationally, has been achieved. In this case, global environmental governance exercised little authority due to the lack of legislative powers to legitimize decisions to its member states.[63] In addition, under the UNFCCC and KP, opposing states are bound by the consensus-based rule.[64]

In general, there was complete consensus on the principles and implementation mechanisms of climate governance represented by the UNFCCC. It was not until 2004 that scientific legitimacy was achieved, acknowledging that there were anthropogenic causes of the problem of global warming. In contrast to the UNFCCC, the KP of 1997 was not designed in a full-consensus format; this resulted in lack of effective governance.[65] Consequently, there were uncertainties that reflected difficulties in the process of decision-making and institutional design.[66]

The KP is a non-consensus mechanism in which different states have different commitments and can compensate for such responsibilities. It encourages

59 Giddens, 2010.

60 Young, 1992:176–92.

61 Keohane, 1998:90–92.

62 Bodansky, 1999.

63 Bodansky, 1999.

64 For a detailed analysis of this legal instrument, see Busby, 2010.

65 Bodansky, 1999.

66 In the late 1960s, J.D. Thompson noted how uncertainties about the problem of climate change can complicate the process of decision-making. He developed a matrix relating beliefs about cause–effect relations and preferences regarding possible outcomes. For further detail see Thompson, 1967.

individual negotiation of GHG management. The KP was seen as a model of "joint intergovernmental cooperation," where production of a global public good (clean air) was assigned to an international organization, but it was based on national cooperation whereby national policies had to adjust and adapt to gain international benefits.[67] In other words, national policies would need to adapt to global needs.

Lack of legitimacy was also expressed in that there was no mechanism with clear sanctioning power to solve disputes.[68] In fact, Article 18 of the KP establishes that the consequences of non-compliance with the Protocol should be expressed only when presenting the consequences and impacts,[69] but there is no mechanism for making it happen.

Yet, to address legitimacy issues and reduce uncertainty, institutions were created. The main structures of the Protocol are the Conferences of the Parties (COPs), established at the UNFCCC. These conferences served as the meeting mechanism of the Parties to the KP. The UNFCCC establishes that

> the Conference of the Parties, as the supreme body of this Convention, shall keep under regular review the implementation of the Convention and any related legal instruments that the Conference of the Parties may adopt, and shall make, within its mandate, the decisions necessary to promote the effective implementation of the Convention.[70]

It was established that COPs should meet in ordinary sessions and could have extraordinary ones. The KP could have observers to the COPs, either states or other entities.[71] In addition, the COPs served as the Conference Meetings of the Parties to the KP (CMP). These conferences were created as a mechanism to include entities other than members of the Protocol. Parties to the Convention that are not Parties to the Protocol may participate in the CMP as observers but without the right to take decisions. The functions of the CMP in the KP are similar to those carried out by the COP for the UNFCCC. The first CMP was held in Montreal, Canada in December 2005, in conjunction with the eleventh session of the COP (COP-11).[72]

Other institutions of the KP were created to help implement the agreement. As stated in Article 6 of the KP, the scope of the Protocol was to prevent emissions from sources and transfer pollution to sinks, generated, acquired or transferred externally. To make this transfer operable, the KP created two permanent subsidiary

---

67 Kaul et al., 2003:11.

68 Bodansky, 1999.

69 United Nations Framework Convention on Climate Change, 1998:15.

70 See Article 7 of the United Nations Framework Convention on Climate Change, 1992.

71 See Article 14 of the KP at United Nations Framework Convention on Climate Change, 1998.

72 United Nations Framework Convention on Climate Change, 2011c.

bodies that also serve at the COPs or CMPs: the Subsidiary Body for Scientific and Technological Advice (SBSTA) and the Subsidiary Body for Implementation (SBI).[73] Subsidiarity implied that the best level for implementing a solution was the one with the competence for so doing.[74] It implicitly meant devolution of authority to the best-equipped problem-solving mechanism or institution. In spite of the political agreements resulting from negotiations, subsidiarity anticipated that there would be situations where scientific expertise was the only alternative to problem-solving or case-reviewing.

In this sense, the SBSTA was established by the UNFCCC to provide the COP and its other subsidiary bodies with timely information and advice on scientific and technological matters relating to the Convention. This body is open to participation of all Parties and is multidisciplinary. It comprises government representatives with competence in the relevant field of expertise.[75] It also develops methodology to define measuring techniques by sector, as well as vulnerability and adaptation techniques. In addition,

> the SBSTA plays an important role as the link between the scientific information provided by expert sources such as the IPCC on the one hand, and the policy-oriented needs of the COP on the other. It works closely with the IPCC, sometimes requesting specific information or reports from it, and also collaborates with other relevant international organizations that share the common objective of sustainable development.[76]

When one country wants, for example, to utilize land-use change and forestry (LULUCF) to account for its GHG reduction, the SBSTA has the competence to decide in which cases this can apply and to add modalities for LULUCF.[77]

On the other hand, the SBI was established to assess and review the effective implementation of the Convention in Article 10(1) of the UNFCCC. An important task is to examine the information in the national communications and emission inventories submitted by Parties in order to assess the Convention's overall effectiveness. In addition, the SBI is in charge of reviewing the financial mechanisms to assist Non-Annex I Parties to implement their Convention commitments.[78] The Global Environmental Fund, administered by the World Bank, is the agency that

73 See Articles 8–10 of the KP at United Nations Framework Convention on Climate Change, 1998:9.

74 Wallace and Wallace, 2000:138–9; Watts, 1999:4–5.

75 See Article 9(1) of the United Nations Framework Convention on Climate Change, 1992.

76 United Nations Framework Convention on Climate Change, 2011a.

77 Although this is a scientific body, negotiations are intense. Some countries, such as Japan and Canada, continue to fight to include widening the sinks to renewable energies or energy savings. The SBSTA has denied this option. See Jaccard et al., 2007:67.

78 United Nations Framework Convention on Climate Change, 2011a.

grants financial aid for KP projects. This is also considered as an operative body for the KP and an essential institution in enabling SBI to accomplish its goal.

In addition, there is a bureau that serves as a representative committee for administrative purposes. There are also several constituted committees for special functions that were created during different COPs. These bodies are the Clean Development Mechanism (CDM) Executive Board for accreditation of operational entities, the Joint Implementation Supervisory Committee (JISC) for verification of emission reduction units (ERU) generated by implementation projects, and the Compliance Committee (CC) for facilitating and enforcing compliance with the Protocol. The KP establishes that Annex I countries must provide information in their national communications to demonstrate that they use these mechanisms as a "supplement to domestic action" to achieve their targets.[79]

Thus the structure of the KP was based on several assumptions, which resulted in rendering impossible achievement of implementation and creating gridlocks. One of the most important assumptions was that countries should measure their emissions on a national basis (absolute emissions). Different countries have contested this form of measurement.

When countries measured their accomplishments based only on the quantity of historical GHGs reduced, there was a significant decrease of GHGs in some countries of Eastern Europe from 1990 to 1995 when their first national inventories were launched. The reason for this decrease was a deceleration of their economies. The deindustrialization and transformation of the economies of these countries helped to achieve very significant GHG reductions. Nonetheless, if we focus our attention on the period after 1995, the decrease has been minimal or zero.

Results for GHG emissions reduction appear worse when a different type of measurement is used. The first option is to measure the intensity factor, that is, the relation between GHG reduction per unit of energy used on a GDP basis. The second is to measure energy used on a per capita basis.[80] If we compare the different types of measurement in relation to sector strategies, the results are quite different and challenge the effectiveness of the absolute-target approach. This difference in ranking of either emerging economies such as China or of the major polluters such as the US is the argument often used to avoid complying with the agreed goals or as a mechanism to negotiate a future agreement. The differences in ranking of intensity, per capita and absolute emissions are shown in Figures 3.2–3.4.

79 United Nations Framework Convention on Climate Change, 2011c.

80 United Nations Environmental Program, 2007:46–7.

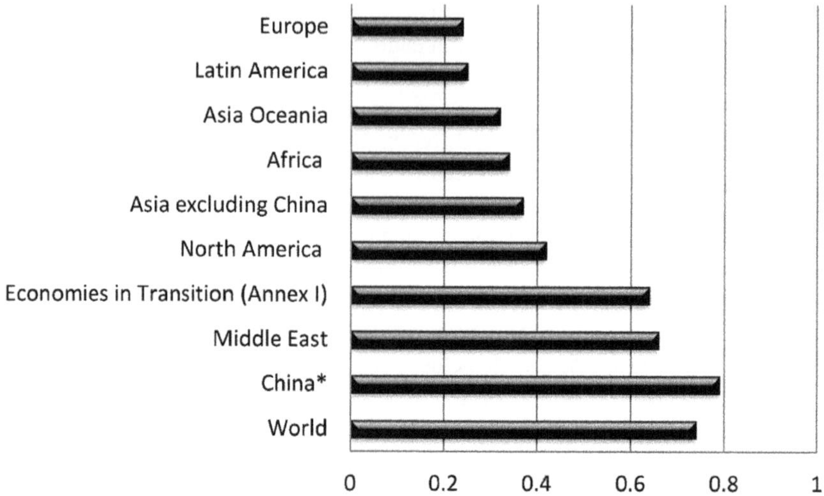

Figure 3.2 Intensity—world GHG emissions in 2010 (MT CO_2/GDP using purchasing power parities)
Source: Adapted from International Energy Agency, 2012a.
* Excluding Hong Kong.

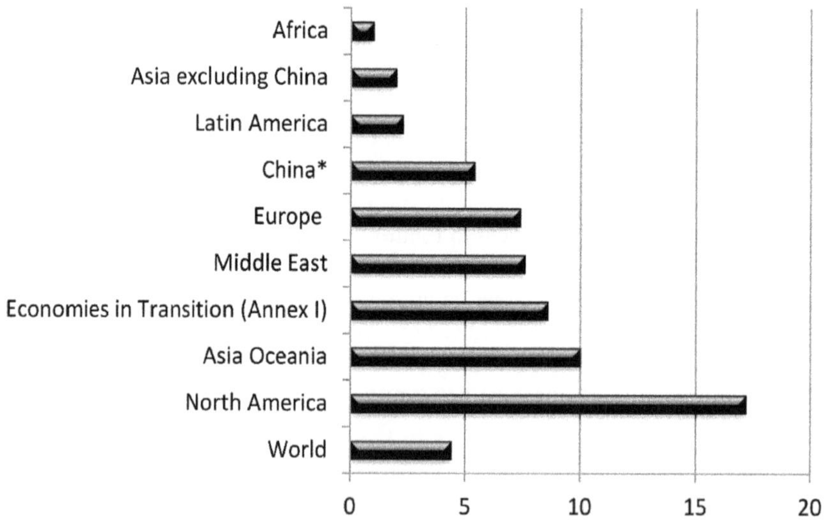

Figure 3.3 Per capita—world GHG emissions to 2010 (MT CO_2 per capita)
Source: Adapted from International Energy Agency, 2012a.
* Excluding Hong Kong.

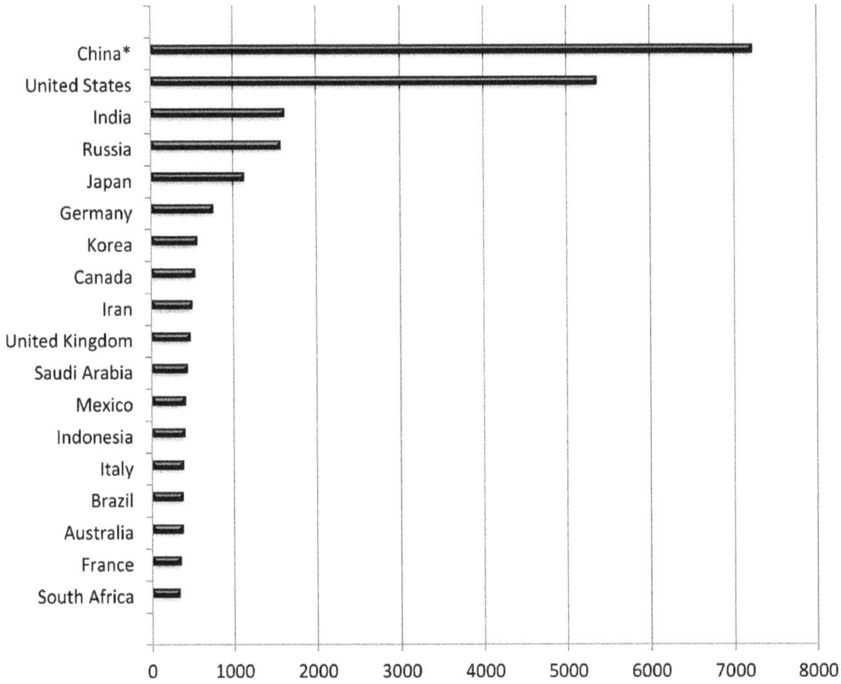

Figure 3.4 Absolute—world GHG major emitters to 2010 (MT)
Source: Adapted from International Energy Agency, 2012a.
* Excluding Hong Kong.

If measured by per capita emissions, the ranking of most polluting countries changes dramatically. For example, if China and Europe take the per capita approach, they appear to have very low levels of GHG emissions, and regions such as North America (especially Canada and the US) would be situated in first place.[81] If oil-producing countries in the Middle East take the intensity approach, they will appear to have very high emissions. Needless to say, this debate is one of the cornerstones of the failure of the KP.

However, it would not be fair to judge the results on a different measurement basis than that established in the KP. Yet it is adequate to measure how effective the KP has been in terms of absolute GHG emission reductions. In spite of the vast institutions created and the potential funding for achieving the goals, GHG total emissions reduction had been modest and far from the goals established. A decade after the coming into force of the KP, almost none of the countries committed under

81 United Nations Framework Convention on Climate Change, 2013a; World Bank, 2011.

Annex I had accomplished their goals. For example, although some countries of the European Union, such as Germany, achieved their Kyoto goals individually, as a region the EU has not. In fact, some other countries, such as Canada and Japan, have increased their emissions. The Energy Information Agency presented data that demonstrate that in 2010 the total level of emissions increased by 3.8 percent.[82]

New big polluters such as the BRICs, South Africa and Mexico have not been doing their job either. Although these countries were registered under a Non-Annex list (which did not imply commitments but only the delivery of annual reports on the state of GHGs to the UN),[83] they register an increasing rate of growth of GHG emissions larger than the majority of Annex I countries. The case of Mexico is presented in Chapter 4.

The GHG increase has to do with the structural framework of the KP, the assumptions and the approaches to implementation. To mitigate this situation, flexible mechanisms were included in the Protocol.

The Three Flexible Mechanisms of the Kyoto Protocol

Annex I countries are required to meet their targets under national regulatory policies. However, the KP provided three complementary market-based mechanisms based on technology transfer and investment, cost efficiency, and on encouraging the private sector and developing countries to contribute to emission reduction efforts.[84]

The three mechanisms are focused on the creation of a "carbon market," where countries that have spare emission units[85] can sell this excess capacity to countries that are not achieving their targets, as established in Article 17 of the KP. In this context, a new commodity was created in the form of emission reductions or removals. Businesses, non-governmental organizations and other legal entities may participate in the three following mechanisms under the authority and responsibility of governments: (1) Emissions Trading Scheme; (2) Clean Development Mechanism; (3) Joint Implementation Mechanism.

The Emissions Trading Scheme (ETS) was established in Article 17 of the KP as an option for individual countries or regions. The idea of setting up this trading scheme was born of a US proposal. However, the Europeans were the ones who developed this proposal through the European ETS, one of the largest

82 Energy Information Administration—US, 2010d.

83 See Article 10 of the KP at United Nations Framework Convention on Climate Change, 1998:10.

84 United Nations Framework Convention on Climate Change, 2011e.

85 Since CO_2 is the principal greenhouse gas, people speak of trading simply in carbon units where Carbon is now traded like any other commodity. There are other units that may be transferred under the scheme, each equal to one tonne of CO_2: (1) a removal unit (RMU) on the basis of land-use change and forestry (LULUCF) activities such as reforestation; (2) an emission reduction unit (ERU) generated by a joint implementation project; (3) a certified emission reduction (CER) generated from a CDM project activity.

operating markets. It was created in 2005 to cover only the CO_2 of energy sources, especially for producing electricity. This ETS granted each European state the right to design their own national plan for allowance allocation, based on Kyoto criteria. It is important to mention that there have been emission reductions due to this mechanism.[86] The proposed Kyoto ETS has not been developed further; it has not operated with the expected success.

The second international transfer mechanism is the Clean Development Mechanism (CDM), which involved investment in sustainable development projects that reduce emissions in developing countries. Under this mechanism,

> (a) Parties not included in Annex I will benefit from project activities resulting in certified emission reductions; and (b) Parties included in Annex I may use the certified emission reductions accruing from such project activities to contribute to compliance with part of their quantified emission limitation and reduction commitments under Article 3, as determined by the Conference of the Parties serving as the meeting of the Parties to this Protocol.[87]

The KP established that this exchange could be between public or private entities and should be voluntary. The CDM has been criticized for rewarding intensive polluters such as China, accused of investing in old-fashioned technology and subsequently being paid to eliminate the source of emissions.[88] The number of projects registered under the CDM reveals that only some countries have had access to this source of funding, as shown in Figure 3.5.

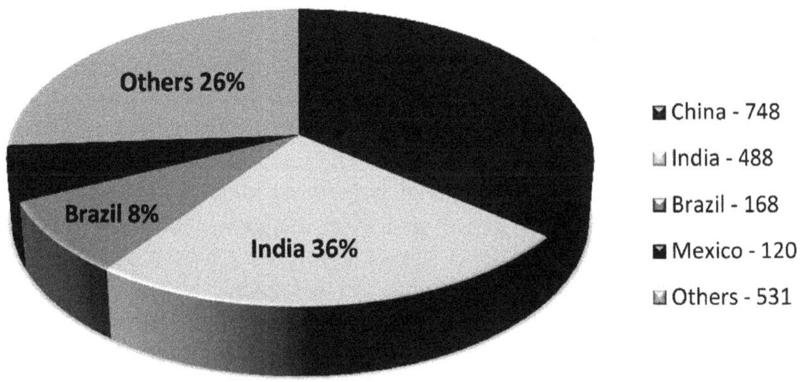

Figure 3.5 Number and percentage of CDM projects by country to 2010
Source: Author's elaboration with data from United Nations Framework Convention on Climate Change, 2010a.

86 Giddens, 2010.

87 See Article 12 of the KP at United Nations Framework Convention on Climate Change, 1998:12.

88 Busby, 2010:7.

Most countries that use the CDM are emerging economies that have been increasing their GHG emissions year after year. Another criticism is the occurrence of free-riding. Some countries that use this mechanism tend to relax their efforts to reduce their emissions. Indeed, some of the CDM projects are perceived as "accounting tricks," not useful at all.[89] However, the CDM has achieved some success, as shown in Figure 3.6.

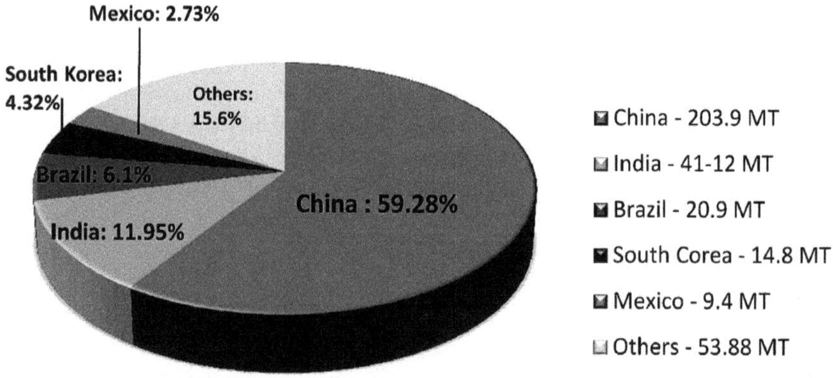

Figure 3.6 Annual average estimates for CO_2e reduction as a result of the CDM to 2012 (millions of tons and percentages)
Source: Author's elaboration with data from United Nations Framework Convention on Climate Change, 2010a.

The third KP instrument is the Joint Implementation Mechanism (JIM), which enables industrialized countries to carry out joint implementation projects with other developed countries. In other words, the transfers are restricted to Annex I countries. As noted by some experts, "a party deciding to host a JI project to generate x tons of CO_2e will have to reduce its assigned amount of GHG by the same tones. Accordingly, it will be 'shooting itself in the foot' if it were to certify an inflated reduction figure."[90] This mechanism has not been very successful either.

In spite of this mechanism, the KP has not performed effective implementation strategies in member countries. In other words, it has not achieved GHG reductions through the JIM. In Article 18, the KP establishes implementation mechanisms to facilitate achievement of goals through the three instruments described above. However, the costs of so doing were localized regionally and dispersed throughout different sectors. Some societies, as in the US, did not want to meet the cost, and politicians did not want to pay the electoral price. In some other countries, such

89 Giddens, 2010:221.

90 Bartsch and Müller, 2000:26.

as Canada, the federal government did not have the competence to implement international agreements dealing with natural resources. This inconsistency between commitments and costs resulted in a lack of effectiveness to implement the international rule.

As shown in this chapter, the KP has had serious flaws from its origins. This situation has driven some countries to set commitments and proposals outside the KP framework. Some groups, such as the G8, the G20, the OECD and APEC, have drafted proposals and positions in which they negotiate under other economic logics and different approaches to measure GHG reductions. In general, these positions are presented at the COPs but negotiated outside of them.

One of these "outside" negotiation forums has been the NAFTA region. This region has taken very different positions on GHG management and experimented with GHG reduction alternatives. Chapter 4 presents the regional and national strategies for the NAFTA area to manage GHGs.

Chapter 4
Formal Regionalism and National Climate Policies in North America

Introduction

Chapter 3 presented the reasons why the global strategy to address climate change had failed in terms of GHG reduction. Some of the reasons had to do with institutional design, relying on a multilateral and national approach, others with lack of capacity for implementation of international institutions. If the global strategy has not fulfilled the need to address climate change, are other levels of governance, such as regional schemes, the key to providing global public goods? Are national governments the appropriate institutions to develop solutions and draft policies in a fully sovereign way? This chapter analyzes how formal regionalism and the governments of the US, Canada and Mexico have addressed climate change in North America. It illustrates how national approaches have had similar results to the global approach. The first section presents the state of the art of the formal regional strategies for climate change in North America. It explains the trinational climate mechanisms working under the framework of the North American integration project, especially within the North American Free Trade Agreement (NAFTA) institutions, the North American Agreement on Environmental Cooperation (NAAEC), the Security and Prosperity Partnership (SPP) and the North American Energy Working Group (NAEWG). Although this formal regionalism has accomplished certain climate-change goals, it has not been enough to start regional cooperation schemes that deal with the provision of global public goods. The second section presents an outline of national efforts regarding climate legislation. Special attention is paid to the US energy transition and the roles of Canada and Mexico in this strategy, as well as the discussion about how "clean" this transition is. The US and Canada have not enacted national climate laws, showing that little attention has been paid to the "cleanness" of the energy transition, while in Mexico the General Law of Climate Change was enacted in 2012, but with very limited implementation capacities. However, as in the case of the formal–regional approach, in none of the three cases have the national governments' policies been effective in reducing GHGs.

Regional Climate Governance: A Trinational Approach

The international trend of environmental protection and GHG emissions reduction discussed in Chapter 3 appeared somewhat inconsistent with the economic dynamics of countries and regions. On the one hand, the fact that the European Union had since the 1990s been deepening its integration with common environmental policies revealed that the UNFCCC's country-based approach would fall short of these new realities. On the other hand, NAFTA played an important role in developing a discourse that tried to link environmental protection with competitiveness.

NAFTA, NAAEC and NAEWG

In the case of North America, regionalism was crystallized by NAFTA in 1994. When reviewing economic integration in the region, it appears that the 1965 Auto-Pact between Canada and the US was the first important agreement of the second half of the twenty-first century. This pact was a free-trade sector agreement that served as a precedent for the automotive chapter in the NAFTA. More than two decades later, in 1989, those countries deepened their integration by signing the Canada–US Free Trade Agreement (CUSFTA). This was an inclusive agreement that integrated sectors such as goods and services, telecommunications and energy sources. CUSFTA was widened when Mexico entered the integration scheme through NAFTA.

Along with NAFTA came the side agreements on labor and environment, as a result of civil-society pressures in the three countries. What detonated the argument and protest against the inclusion of these agreements were the lawsuits brought to the US and Mexican governments by the NGO Public Citizen and other national environment organizations for not incorporating environmental issues in NAFTA.[1] Although these suits were not accepted by either country's courts, they established the conditions for collective contentious action. Environmental groups and non-governmental organizations such as Friends of the Earth, the Sierra Club, Greenpeace, the World Wildlife Fund and The Nature Conservancy, among others, pressured legislators and executives (especially in the US) to include environmental topics in the trade agreement.[2] The result was the enforcement of the North American Agreement on Environmental Cooperation (NAAEC), yet with no direct mechanisms for solving controversies linked to NAFTA. The resulting institution was a mechanism to appease environmental groups concerned about the ecological impacts of increased trade in North America.[3] This situation confirmed that trade took precedence over any environmental pledge.

1 Kibel, 2000:12.

2 Torres, 2009:198.

3 Betsill, 2006:37; Mumme, 1999:1.

This explains the NAAEC's institutional limits regarding its state-centered nature, which prevented it from constructing mechanisms that would generate a self-sustaining dynamic and effectiveness at the continental level.[4] Alberta Sbragia points out that NAFTA was designed to avoid institutionalization and institution-building in a supranational approach. The trade agreement was based on the national states and exemplified by the rules of origin and the independence of each nation's trade policies. Nonetheless, although no supranational institutionalization was crafted, the North American project created a dense network of relations, contacts and cooperation mechanisms in which institutional relations were regularized.[5] In addition, the NAAEC clearly established that each of the member states would establish national laws to protect the environment, and that sovereignty would remain the fundamental principle for intergovernmental cooperation. Nonetheless, in spite of its weakness, the NAAEC has been working with a trilateral network of actors (policy-makers, epistemic communities, business, local governments).

Originally, the NAAEC was an agreement whose main goal has been "to address regional environmental concerns, help prevent potential trade and environmental conflicts, and to promote the effective enforcement of environmental laws."[6] The NAAEC works to develop trilateral projects and research through the Commission for Environmental Cooperation (CEC) and deals with environmental problems resulting from trade in the three countries. However, climate change was by no means a priority issue for CEC when the North American project was designed.[7]

Although the NAAEC does not address the GHG problem directly, it was supposed to become a platform from which governments and civil society could ask for aid and denounce environmental violations. Thus the institution would favor not only state-to-state cooperation, as intended, but also transnational and transborder links.[8]

Other kinds of limitations of the NAAEC are technical, such as the fact that the panels for solving controversies "are not entitled to review the adequacy of laws, nor any amendment to them, as is the case within Chapter 19 of the NAFTA."[9] Likewise, the activation of a panel is not automatic, as in the rest of NAFTA's mechanisms; during the research period the Council of the NAAEC can decide to stop the case.[10] In addition, the preamble of the agreement left environmental

4 Clarkson, 2010:290.

5 Sbragia, 2010:268–73.

6 Commission for Environmental Cooperation, 2009.

7 Betsill, 2006:37.

8 Stephen Clarkson points out that in the North American project some economic sectors have demonstrated significant transborder governance (steel, automobiles, textiles); in some others, harmonization has resulted more from global governance influence (intellectual property rights for pharmaceuticals, financial services). For further details, see Clarkson, 2010.

9 Morales, 2008:53.

10 See Article 24 of the North American Agreement on Environmental Cooperation, 1994; see also Morales, 2008:54.

protection to the jurisdiction of national governments, making the possibility of developing regional environmental standards, for example climate-change emissions inventories, very low.

The NAAEC created the Border Environment Cooperation Commission (BECC), the North American Development Bank (NADBank) and the CEC, and a funding agency called the North American Fund for Environmental Cooperation (NAFEC). These institutions worked as the framework for cooperation on environmental trinational projects. Although several projects are still operating, the funding agency worked only from 1996 to 2003.[11] Thus "there continues to be a mismatch between the depth of the region's environmental problems and the resources devoted to mitigate them."[12]

The advantages of having these institutions are that, in spite of lacking a binding mandate in the three countries, they can have a positive influence on environmental protection.[13] For example, the CEC's main areas of performance when it was created were biodiversity conservancy, pollutants and health, sustainability and trade. Having jurisdictional limitations, its most important goal was the provision of environmental information on each of these issues. The CEC became the forum for "sharing data and analysis, and to boost the quality, comparability and compatibility of national and subregional information."[14] This is the case for the topic of climate change, which was included in 2010 as part of a five-year strategic priority. The three projects under this topic are aimed at compiling information about carbon storage, emissions and inventories. The goal is for the three countries to have a comparable database regarding climate change and to be able to share an interactive informational platform. However, there are no projects designed to intervene directly with the emissions reduction path of any of the three countries. The CEC has served as an information compiler for GHG data of the three countries, but not as a means of policy-making design or implementation. Nonetheless, it has been an important partner when addressing methodological problems of measuring GHGs and designing databases. These problems come in various forms. First, there are different jurisdictional types that measure GHGs: some are national (in the case of Mexico there is the Registro de Emisiones y Transferencia de Contaminantes); others are local (as in the US and Canada, where registries are designed by province or state); and yet others follow the methodology of global institutions such as the KP. This project intends to harmonize and diffuse this basic element of climate-change mitigation and adaptation "with the reporting guidelines developed by the UNFCCC."[15]

The area in which the CEC has performed consistently is border environmental cooperation between the US and Mexico. As mentioned above, the NAAEC created

11 Commission for Environmental Cooperation, 2009.
12 Schott, 2008:1–2.
13 Kibel, 2000:15.
14 Commission for Environmental Cooperation, 2009.
15 Commission for Environmental Cooperation, 2013.

the Border Environment Cooperation Commission (BECC), which is responsible for several projects related to climate change. The first one addresses the management of methane sources and their transformation into biogas. This program is operated in coordination with the Environmental Protection Agency (EPA). The second is a program to help local governments to develop their climate action plans. Thanks to the BECC, northern Mexican states have accessed EPA methodologies for drafting their emission inventories, standardized with those of the US states and Canadian provinces. For example, the GHG emissions inventory of the state of Chihuahua drafted in October 2010 received funding and support from the BECC, and the Center for Climate Strategies helped in its design, together with the state's government.[16] In the NAFTA context, the harmonization of methodologies for measuring GHGs becomes important, due to the fact that production chains are integrated. However, these methodologies are dissimilar from those proposed by the KP, a situation that presents a problem and forces local governments (LGs) to choose between the methodologies to develop their inventories.

Driven by NAFTA, the energy sector also addressed the issue of climate change, albeit tangentially. NAFTA was perceived as an opportunity to create a truly integrated regional market to reduce energy dependence on the Organization of Petroleum Exporting Countries (OPEC). One of the interests of the US in negotiating NAFTA was to create a free-trade area for the energy sector, where the Mexican energy monopolies could open themselves to trade.[17] The US and Canada had already established free trade in this sector via CUSFTA,[18] but Mexico was not willing to open in this sector and presented constitutional arguments against regional openness.[19] It is worth noting that, in spite of the resistance of Mexico to being part of a regional free market for energy, the country had a long trading history with the US. Pushing forward this attempt to create an energy-integrated regional market, in 2001 the three leading executives, through the Secretary for Energy in the US, the Secretary for Energy in Mexico and the Canadian Minister for Natural Resources, created the North American Energy Working Group (NAEWG). This would serve to improve energy trade and interconnections within North America.[20]

16 Chacón et al., 2010.

17 Puyana, 2006.

18 The rules established in CUSFTA were consistent with those of the General Agreement on Tariffs and Trade (GATT), later the World Trade Organization (WTO).

19 Chapter VI of NAFTA (dealing with the energy sector) was negotiated differently with Mexico than between the US and Canada. For example, it did not apply to five areas in the Mexican energy sector: exploration, exploitation, production of oil and natural gas, petro-chemical services and the guarantee to satisfy domestic (regional) demand. For further details, see Puyana, 2006; Domínguez, 2011.

20 See De la Vega, 2006; Department of Energy—US, 2007. It is important to note that the NAEWG was elaborated on the basis of the "Cheney Report" of 2001. This document established the need to invest in transnational infrastructure (for fossil fuels and electricity). For further information, see National Energy Policy Development Group, 2011.

Some years later, the NAEWG became part of a greater initiative, the Security and Prosperity Partnership for North America (SPP), signed in 2005 by the executives of the US, Canada and Mexico. The SPP focused on several areas, such as energy, transportation, financial services, technology, business, environment, agriculture and health.[21] The goals for the areas of energy and environment were very similar to those of the NAEWG. The SPP's aim was to help develop frameworks to harmonize energy efficiency standards, boost North American energy integration, develop clean energy technologies and improve North America's air quality and safety in the marketplace.[22] The SPP energy group encouraged energy interconnection in the region mainly through the major federal energy regulators in North America, the US Federal Energy Regulatory Commission (FERC), the Canadian National Energy Board (NEB) and the Mexican Regulatory Energy Commission (CRE).[23]

The SPP created different working groups, and decided to incorporate the already-functioning NAEWG to continue to develop regional energy integration. The NAEWG/SPP, to date, is composed of nine groups of experts dealing with energy efficiency, technology and alternative clean energy sources.[24] There are several areas in which the NAEWG/SPP has performed energy integration, as shown in Table 4.1.

Table 4.1shows that energy integration has moved in the direction of greater energy efficiency, where most achievements have taken place. When dealing with climate change, there are only two indirect efforts regarding alternative energy production: tidal energy and biofuels. The only direct initiative regarding climate change was carbon capture and storage (CCS). As a result of the creation of the North American Carbon Atlas Partnership in 2009 (part of the NAEWG project), the three countries developed a North American CCS map with a common methodology and the goal of sharing and comparing data.[25] Yet, the CCS technology has not been tested in large-scale projects (once CO_2 is pumped underground, there is no guarantee that it will not leak).[26] This was a very clear message that the North American energy integration scheme, driven by SPP, focused on keeping up with economic growth, saving energy, and finding fuel substitutes. As Morales notes,

21 Department of Energy—US, 2007.

22 Villareal and Lake, 2009:2; Department of Energy—US, 2007.

23 National Energy Board—Canada, 2009a:21.

24 The working groups are: (1) Electricity Experts Group; (2) Energy Efficiency Experts Group; (3) Energy Picture Experts Group; (4) Hydrocarbons Experts Group; (5) Natural Gas Trade and Interconnections Experts Group; (6) Nuclear Collaboration Experts Group; (7) Oil Sands Experts Group; (8) Regulatory Experts Group; (9) Science and Technology Experts Group.

25 For further details, see North America Energy Working Group, 2012.

26 Carbon capture and storage (CCS) involves capturing CO_2 produced during the process of power generation and storing it permanently in deep geological structures beneath the land or sea. Plymouth Marine Laboratory, 2013; Clarke, 2008:156.

Table 4.1 Energy groups in NAEWG

Area of energy integration	Status
Energy efficiency standards	Three energy performance standards have been harmonized across the three nations: freezers and refrigerators; three-phase motors; and room air conditioners. There is a commitment to further align energy efficiency standards on key consumer products; and agreements to identify specific ways to increase cooperation on research and development and reduce barriers to deployment of new technologies in a wide variety of areas, including biofuels, gas hydrates, hydrogen, carbon capture and storage, clean coal and electricity transmission.
Liquefied natural gas	There are efforts to explore ways to strengthen the market for LNG.
Joint modeling effort	A trilateral experts group has created a model that offers a supply and demand energy outlook to 2030. The forecast maps out the future of electricity generation, oil and gas production and consumption, and coal supply and demand in each of these three countries
Electricity networks	Discussions have begun on smart grid cooperation and other transmission-related issues. Further efforts include possible joint work on the effective implementation of intermittent renewable generation technologies into the grid, including load management and grid stability.
Ocean energy	There are efforts to harness ocean energy for electricity generation.
Biofuels for transportation	There is the objective to build a North American market for biofuels
Carbon capture and storage (CCS)	The creation of the North American Carbon Atlas Partnership in 2009 was directed to develop a North American CCS map.

Source: North America Energy Working Group, 2012; Department of Energy—US, 2007.

these goals fit perfectly into the US agenda of the George W. Bush administration.[27] The role of the other two countries appears to be directed to keep both Canada and Mexico as reliable and safe suppliers of the US.[28]

27 Morales, 2008:172.

28 Clarkson and Mildenberger, 2011; Morales, 2008:173.

However, it seems that, with the administrations of Barack Obama, Stephen Harper and Felipe Calderón, a strong link between energy efficiency and GHG reduction was consolidated. In late August of 2009, the three executives drafted a joint declaration on climate change in which practical goals were set to create conditions for low-carbon growth plans. This declaration went beyond former objectives: it pleaded for the harmonization of measurements, reports and verification of GHG reduction, and recommended that certain sectors (such as oil, gas, transportation) enter the low-carbon economy schemes. This effort would be accompanied by funding, clean technology and infrastructure, especially for emissions trading.[29] Thus it could be assumed that the trend for energy transition in North America would also support the integration of climate-change policies.

As mentioned earlier, it is quite clear that the leader of the North American transition has been the US, and, as some authors note, the strategy has been to consolidate two bilateral energy relations (US–Canada and US–Mexico) that constitute a regional transition. An energy transition can be understood as a significant set of changes to the patterns of energy used in a society. It is driven by new resources and technology supported by institutional capacities derived from policy-making that meets energy demand for services at low cost. Peter O'Connor proposes four elements of energy transition: the change of (1) energy resources (e.g. incorporation of shale gas); (2) energy carriers (e.g. electricity or gasoline); (3) energy converters (e.g. light bulbs or automobiles); and (4) energy services.[30] For example, the US lived through an energy resource transition—from 1870, when 70 percent of its industry used wood, to 1990, when this was substituted by coal, and to 1960, when coal was substituted by oil and gas. My view is that the energy transition of the US in the early 2000s is based on energy resources and energy services, elements needed to address two main constraints: climate change and limits to oil supply.[31]

Energy transition in North America implies changing or substituting traditional fossil fuels for forms of non-conventional or cleaner forms of fuel. Since the 1990s the US energy transition has rested on these elements, aimed at self-sufficiency facilitated by innovation in energy carriers and converters. As explained by Rhead Enion, the challenge of the energy transition in North America is to reduce coal and oil as the main energy sources (especially for heavy vehicles and transportation); the transition in carriers such as electricity would be easier to achieve once the sources changed.[32] In this sense, energy transition in the region implies addressing climate change directly with mitigating strategies such as change of energy sources and converters.

However, this transition has not been as successful as expected. Data show that North America has a 5.13 regional average when relating CO_2e and electricity

29 The White House, 2009.

30 O'Connor, 2010.

31 Ibid.; Gibbons and Blair, 1991.

32 Enion, 2013.

production, a very high figure when compared with Europe's 3.18 average (OECD countries) and 3.75 of the Asia Pacific (OECD countries). However, if compared with some non-OECD countries such as China (8.30), India (8.41) or Saudi Arabia (6.09), the region is better positioned. The US and Mexico contribute most, with 5.5 and 5 respectively, compared with Canada, which accounts for only 1.8, thanks to the high-performing hydropower sector.[33] In the US and Canada, production of electricity, transportation and industry account for more than 80 percent of GHG emissions;[34] in Mexico, 60 percent of GHGs come from the energy sector.[35]

GHGs and the Energy Sector in North America

As mentioned earlier, in the three countries of North America, the source that emits most GHGs is undoubtedly production and consumption of energy. Therefore regional GHG management needs to be connected to the regional energy transition to include new energy sources (non-conventional or renewable). But new sources are not enough in a region that has profound commercial integration. Physical interconnection is fundamental to this transition. The infrastructure consists of (1) power plants for electricity generation and transmission systems, (2) pipeline systems for natural gas, gas storage and liquefied natural gas facilities and (3) refineries, pipelines and import facilities for oil.[36] Significant facilities have been constructed to interconnect the exchange of energy sources. Up to 2005 there were 35 cross-border natural gas pipelines, 22 cross-border oil and petroleum product pipelines, and 51 cross-border electricity transmission lines connecting the three nations.[37] Between 2008 and 2012, the number of facilities increased to 49 border gas pipelines and 8 maritime ones. The facilities for providing oil and petroleum products to North America have also increased to 65, and 10 more cross-border electricity projects have been completed.[38] In spite of the efforts of the US to create a common energy market, the reluctance of Mexico to enter this scheme resulted in two bilateral relations, that between the US and Canada, based on a free-trade approach, and that between the US and Mexico, with a restrictive approach.[39]

Historically, the US has been the main importer of crude oil and petroleum products, much of which comes from Canada and Mexico.[40] Since the 1970s, the commercial relation in the region has deepened, representing a growing exchange

33 Secretaría de Energía, 2011a. See original data at International Energy Agency, 2008.

34 Environmental Protection Agency—US, 2008.

35 Secretaría de Medio Ambiente y Recursos Naturales, 2009b.

36 Prado, 2007:406.

37 Ibid.:407.

38 Lum and Knutson, 2013; Energy Information Administration—US, 2008b.

39 Barnes, 2011.

40 Crude oil represents raw unprocessed resources; petroleum products imply gasoline, asphalt and chemical products.

of these products year by year. Stephen Clarkson points out that, while the US is the third-largest crude oil producer, after only Saudi Arabia and Russia, its domestic production satisfied only 51 percent of its needs for 2011.[41] Nonetheless, according to scenarios of the 2012 World Energy Outlook, the US will become the number one oil and gas producer in 2025. This will be possible thanks to technological improvement in respect of new energy sources, such as light tight oil and shale gas. This report seems to be optimistic about the prospects for this path to result in fuel efficiency and in the decrease of oil imports by the US, to become a net oil exporter around 2030 (especially to Asia).[42]

According to the US Energy Information Administration, in 2011, the region accounted for 15 percent of the world's proven oil reserves. Canada reported almost 13 percent of oil (most of it from non-conventional sources such as tar sands), the US 1.4 percent of crude oil, while Mexico reported only 0.77 percent, also of crude oil.[43] It is worth noting that the US and Canada are and will be the major producers of these fuels up to 2035, with 6.6 billion barrels.[44] Canadians and Americans share the closest energy relation in the world. Canada is the major supplier of energy to the US in terms of crude oil and refined petroleum products, natural gas, electricity, coal and uranium. In 2011, Canada's energy exports were valued at 120 billion USD (119 billion CAD), with virtually all (90 percent) of it going to the US. In addition, Canada exported 2.7 million barrels per day of crude oil and refined products to the US, representing 24 percent of total US petroleum imports, supplied approximately 20 percent of the uranium used in US nuclear power plants, and provided 90 percent of all US natural gas imports, representing 13 percent of US consumption; it imported 55 billion CAD of energy products, of which 33 percent came from US–Canadian natural gas imports, which now stand at almost 3 billion cubic feet per day and have approximately tripled since 2006. With the exception of very small amounts of natural gas imports, Canada purchases most of its natural gas from the US. In the electricity sector, the Canadian and US electricity grid is deeply integrated, with more than 30 major transmission lines connecting almost all Canadian provinces to neighboring US states.[45]

In the other bilateral energy relation, Mexico–US, we find that total energy exports of Petróleos Mexicanos (the state-owned monopoly company) increased from 38,000 million USD in 2006 to 51,000 million USD in 2011.[46] Mexico is a world-class oil producer with 2.5 million barrels daily (mbl/d), of which 1.2 are exported primarily to the US (1 mbl/d), which represents more than 10 percent of its consumption, to Spain (110,000 bbl/d), India (37,000 bbl/d), China (36,000 bbl/d), Canada (20,000 bbl/d), and to Central America and the Caribbean, through

41 Clarkson and Mildenberger, 2011:60.

42 International Energy Agency, 2012b.

43 Energy Information Administration—US, 2012a.

44 Secretaría de Energía, 2012b:40.

45 Government of Canada, 2012.

46 Secretaría de Energía, 2012a.

the San José Agreement.[47] North America accounted for 217.5 bbl/d of proven reserves up to 2011, representing a total of 13.2 percent of world reserves. Within the region, Canada accounts for 80 percent, the US for 14.2 percent and Mexico for 5.8 percent.[48] Regarding production, the numbers change slightly, with the US producing 54.8 percent of the region's oil, followed by Canada with 24.6 percent, and Mexico with 20.5 percent up to 2011.[49]

The relation in North America has been based on cooperation, especially since the energy transition started. During the first decade of the 2000s, the US and Canada have advanced in cooperation schemes to integrate more efficiently the exchange of fuels. One of the most discussed projects recently is the Keystone Pipeline, which would connect Alberta to Texas to provide fuel. This project is under analysis by the Obama government and has resulted in serious dissent in both states. We should remember that, since his second period in office, Obama's agenda has included fighting climate change; it is likely that the Keystone Pipeline will face delay during his administration. In this regard, the president pointed out that "the net effects of the pipeline's impact on the climate will be absolutely critical to deciding whether this project goes forward."[50]

Between Mexico and the US, relations have also deepened lately. In 2012, an agreement between Mexico and the US regarding transboundary oilfields in the Gulf of Mexico was signed. This treaty will establish a cooperation framework to exploit shared hydrocarbons in the Gulf of Mexico.[51]

Apart from oil, natural gas represents the second major fuel traded in North America. The trade relation in the region is *sui generis*. Canada exports go directly to the US, while the US produces gas for domestic consumption and sells part of it to Mexico, which is the largest importer in the region.[52] This relation will probably change due to the inclusion of shale gas in the three countries' energy mix. According to data from the International Energy Agency (IEA), the US will become a super-exporting power of shale gas; Canada and Mexico have also recently discovered vast reserves of this gas.

One major part of the fuels traded in the region goes to electricity generation and to transportation. In general, the electricity sector is as integrated as the fossil fuels sector. Integration is deep between Canada and the US. Both countries have

47 The San José Agreement includes Barbados, Belize, Costa Rica, El Salvador, Guatemala, Haiti, Honduras, Jamaica, Nicaragua, Panama and the Dominican Republic. Exports to China started in 2010. Both India and China have registered an increasing demand for Mexican oil since 2006. Data for these countries can be found at Secretaría de Energía, 2012a: 99.

48 Secretaría de Energía, 2012b:22.

49 Ibid.:28.

50 Rincon, 2013.

51 At the time or writing, this agreement was still waiting for ratification by the US Congress.

52 Energy Information Administration—US, 2012a.

four major transmission grids that provide power inter-regionally—the Western Interconnection, the Eastern Interconnection, the Electric Reliability Council of Texas and the Province of Quebec—with about 204,000 miles of long-distance transmission lines and about 157,810 miles of transmission lines within the US.[53]

Certain states and provinces in both countries integrate these electricity grids. For example, the Western Electricity Coordinating Council (Western Interconnection) integrates western US states, British Columbia, Alberta and the Mexican state of Baja California.[54] The Midwest Reliability Organization[55] includes US states, Saskatchewan and Manitoba; the Northeast Power Coordinating Council[56] (Eastern Interconnection) represents the East Coast US states as well as Ontario and Quebec.[57] The relation between Canadian provinces and US states proves to be regionalized. For example, "many provinces currently have stronger interconnections in a north–south direction in order to allow for lucrative electricity trade with the United States, rather than in an east–westerly direction that would allow for a pan-Canadian electricity market to emerge."[58] The energy trade is vertical and regional, and tends to flow south. It is worth noting that

> net interregional trade accounted for less than 1 per cent of delivered power in 2010 … Electricity flows south from the Northwest to California and the Southwest. It flows south from Manitoba to the Midwest and from there to the Central region (the Southwest Power Pool) and the Tennessee Valley Authority (TVA). Electricity also flows south from TVA to the Southern region (Southern Company) and from the Southern region into Florida. It flows south from eastern Canada into New England, New York, and the Midwest.[59]

As shown in Figure 4.1, the interconnection lines that go south represent the majority of the transmission lines of the North American electricity relation, especially between Canada and the US.

In the same vein, Canada has historically been the largest exporter of oil (extracted from tar sands) and natural gas to the US.[60] However, these reserves are

---

53 Prado, 2007:409.

54 For a detailed list of members, see Western Electricity Coordinating Council, 2011.

55 See Midwest Reliability Organization, 2011 for a detailed description of the organization.

56 In 2007, the Northeast Power Coordinating Council—Cross Border Regional Entity, Inc.—was merged with and into the Northeast Power Coordinating Council, Inc. The resultant Council, as the surviving entity of the merger, is the cross-border regional entity and services corporation for Northeastern North America. For a detailed list of members, see Northeast Power Coordinating Council, 2011.

57 Morales, 2008:178.

58 Canadian Electricity Association, 2006:14.

59 Energy Information Administration—US, 2013a.

60 Especially since CUSFTA and then NAFTA, which established the "proportionality clause" for Canadian energy sources that forbids the country from reducing exports to the

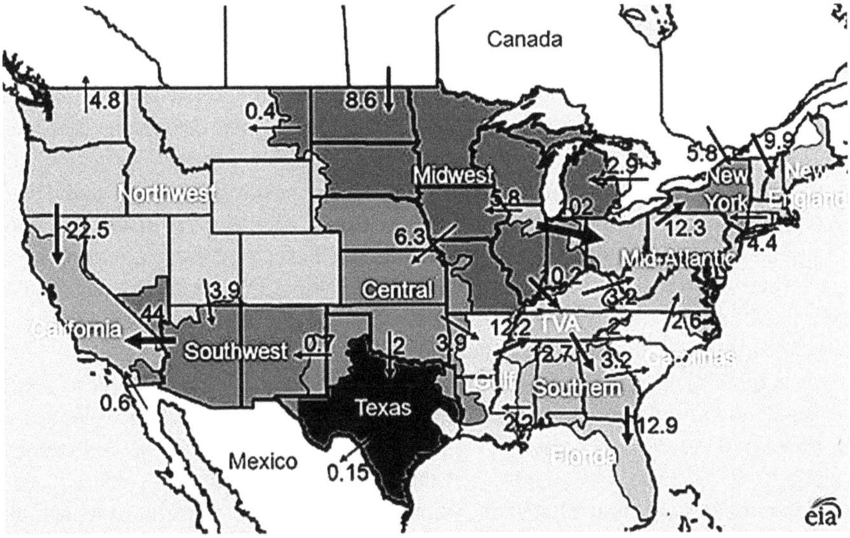

Figure 4.1 North–south electricity connection in North America, 2010 (million MWh)

Source: Energy Information Administration—US, 2013a.

concentrated in Alberta (the largest producer), British Columbia, Saskatchewan, Newfoundland and Nova Scotia.[61] Infrastructure to transport both fuels runs from these provinces throughout the US states. Pipelines, such as the Alliance Pipeline, the TransCanada Pipeline System and the West Coast Energy pipelines represent this large-scale integration of infrastructure.[62] Also at a transregional level, infrastructure has been developed with the goal of creating 80,000 km of connecting pipelines that run from British Columbia through Idaho to California. The connecting infrastructure that provides the region of the Great Lakes and the East Coast runs a 50,000 km pipeline system.[63]

Even though the relation is vertical, it is so entangled that, when it is cold in eastern Canada and in the eastern US, producers in British Columbia can swing their production to pipelines (either in Canada or through the US) heading east and take advantage of higher market prices in that market. When the market demands more gas in California, the gas can be redirected to the south, or, when Vancouver gets unexpected snowfall and gas demand increases, gas can be switched to yet another pipeline system, to be delivered to the lower mainland. Similarly, natural

US because of prevention policies relating to preserving natural resources. See Vargas and Hickman, 2010.

61 Prado, 2007:415.

62 Ibid.

63 Ibid.

gas buyers who are connected to different sources of natural gas or oil can source their supplies from different markets, depending on their needs.[64]

Oil infrastructure (crude and refined) is also heavily integrated between the US and Canada. In a horizontal scheme, pipelines connect major Canadian cities and the Great Lakes area. Vertical oil integration advocates a complex infrastructure system that connects Canadian province suppliers, such as Alberta and New Brunswick (crude and refined) with the East Coast, central parts of the US, and Texas, one of the most important suppliers. On the West Coast the major producers are located in Alaska, Alberta and California. Their infrastructure is developed near these centers, running from Alberta all through the West Coast to California, also integrating Illinois into the system.[65]

It is quite clear that the US and Canada share an energy market. But in spite of the efforts of the US to create a North American energy market, the reluctance of Mexico to enter this scheme has resulted in the creation of two bilateral energy relations. In Mexico, production, distribution and management of oil and gas resources have been controlled by the state-owned monopoly Petróleos Mexicanos (PEMEX) since 1934. PEMEX is one of the largest producers of oil worldwide, and its network for transporting oil (crude and refined) and gas consists of almost 47,000 km. Of the total of PEMEX oil exports, 90.7 percent is destined to the US.[66] Besides controlling all the oil and gas infrastructure in the country, PEMEX "also controls 50 percent of the 334-thousand-bbl/d Deer Park Refinery in Texas,"[67] showing that the relation with the US is fundamental to providing energy (especially petroleum and its products) to the Mexican population.

Mexico has a solid relation with the US regarding the oil and natural gas trade, being a net importer of gas and oil fuels and an exporter of crude oil to that country.[68] Up to 2009, the country had been the seventh worldwide oil producer, the sixteenth exporter and the eleventh oil consumer.[69] The majority of its exports go to the US, being in the top three suppliers since the year 2000. Regarding the gas sector, Mexico imports gas to supply its growing domestic demand. In comparison with the other two partners, Mexico is the most "addicted" to gas.[70] It has 15 interconnections with the US, of which seven can carry gas bi-directionally to south Texas.[71] There are some transboundary projects to be built soon: two pipelines between Texas and the Mexican border states of Nuevo León

64 National Energy Board—Canada, 2009a:7.

65 Prado, 2007:414.

66 Secretaría de Energía, 2007b:86.

67 Prado, 2007:417. Bbl/d stands for barrels per day. One barrel of oil represents 42 US gallons or 158.9873 liters.

68 Vargas and Hickman, 2010.

69 Barnes, 2011.

70 Morales, 2011. See also Vargas and Hickman, 2010.

71 Secretaría de Energía, 2007a.

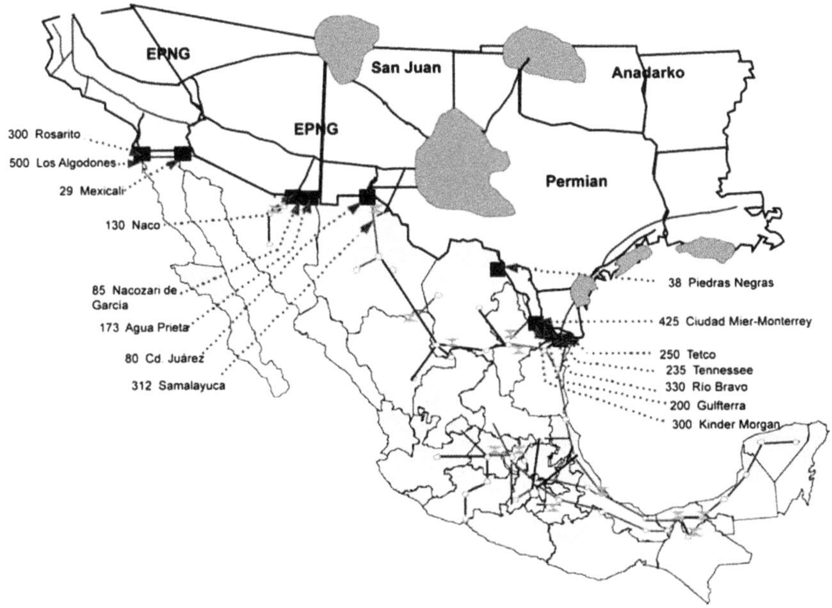

Figure 4.2 Natural gas interconnection points between Mexico and the US
Source: Secretaría de Energía, 2007a.

and Coahuila.[72] Figure 4.2 shows the main interconnecting points of natural gas between the two countries.

The electricity sector in Mexico is also managed by the state-owned company Federal Commission of Electricity (CFE). In the West Coast area, electricity has been exported to CFE through a system of transmission lines located in Mexicali, Baja California and Sonora.[73] In other parts of the border, there are also interconnections that provide electricity to the northern states of Mexico, through contracts between US companies and CFE. These arrangements incorporate indirectly the northern states of Mexico into the already established electricity regions in the US, such as the Western Electricity Coordinating Council (WECC) or the Electric Reliability Council of Texas (ERCOT). In the trade relation between US and Mexico, the latter country has had a surplus since 2003.[74]

In general, it is important to note that energy trade between Canada and the US is based on oil and gas markets, and depends on the efficiency of the operation and infrastructure. The physical connection among energy partners speaks of vertical integration and regionalization processes, not between the three countries,

72 Prado, 2007:409–17.
73 Secretaría de Energía, 2007c:47.
74 Energy Information Administration—US, 2011.

but among neighboring local governments. Although there is a thin line between what is regional in national terms and regional in transnational terms regarding energy policies, the interconnection of energy infrastructure among certain areas of Canada and the US is solid and integrated. In this regard, mechanisms under regional trinational schemes have succeeded in integrating energy markets of conventional and non-conventional sources, but they have failed to find solutions to the problem of climate change resulting from energy use and production in the NAFTA context.[75]

National Climate and Energy Strategies in North America

In spite of having environmental trinational institutions created within the commercial integration process, North America has shown very little regional coherence regarding climate change. It was expected that a highly integrated region would act as the EU did, and perform aggregate efforts to contribute to GHG reduction and the provision of global public goods. However, this was not the case. The three North American countries addressed climate change using different approaches. The US did not ratify the KP; Canada committed to reduce 6 percent of GHGs; Mexico as a Non-Annex country had no reduction commitments, but ratified the KP. Each of these countries calculated the costs and benefits of ratifying and implementing the KP. The US Congress decided not to enter a process that would affect its economy; Mexico ratified because it would have no reduction commitments and could benefit from the international mechanisms. However, Canada ratified the KP in spite of knowing that the costs would be very high. Harrison and McIntosh argue that countries in this position (the EU, Japan, Australia and Canada) entered this dynamic either because of voters' pressure or because of politicians' own beliefs; Canada is an example of both.[76]

Nonetheless, the three countries have not been completely successful in addressing climate change. It should be noted that the sole Non-Annex country in the region, Mexico, is the only one to have enacted a national climate-change law; the other two countries have different regulations but have failed to create a national climate-change standard to help coordinate local climate policies. On the other hand, it is significant that the only country not historically part of the KP in the region and with no general law, the US, has had better GHG reduction results than the others. This leaves Canada in the most unfavorable position, becoming even worse when considering the withdrawal of that country from the KP in 2011.

To understand the politics and policies around the UNFCCC and the KP, this section analyzes the national debates around climate change and the legislative outcomes.

75 Winfield et al., 2008.

76 Harrison and McIntosh, 2010.

Canada: Climate and Energy National Legislation

The topic of climate change has been on Canada's agenda since the 1980s. The first efforts of the country to deal with air pollution were the Vienna Convention of 1985 and the Montreal Protocol of 1987. Both instruments addressed the depletion of the ozone layer by banning specific industrial polluting substances. The Convention, as well as the Protocol, served as a basis for the creation of the IPCC in 1988 and the development of the international conference "The Changing Atmosphere: Implications for Global Security," held in Toronto that same year.[77] This environmental inertia put the topic of climate change onto the Canadian agenda, which positioned the country as a global leader for the Rio Conference and the creation of the UNFCCC.

In 1995, Canada drafted its first climate action plan, called "Canada's National Action Program on Climate Change." This document provided a starting point for the negotiations among society and provincial governments to pursue a common position facing COP-3 in Tokyo. Canada ratified the KP in December of 2002 and committed to 6 percent GHG reduction. It is important to mention here the pressure put by civil society for this ratification. The Pembina Institute, the David Suzuki Foundation and the Sierra Club organized national campaigns to accomplish this.

Since then, the federal government has presented different documents to implement the KP commitments. As noted by Harrison, it encouraged the business community and provincial governments to adopt relevant measures. The main incentive used back then was public spending, not regulation.[78] For example, there was the Climate Change Plan for Canada of 2002 and Project Green in 2005. Unfortunately, these plans were not successful, and the country continued to increase its GHGs. In addition, these federal plans generated conflict with some provinces. Regional economic integration among provinces and states of the US meant that "Canadian manufacturers feared not only that the costs of compliance with Kyoto would place their goods at a competitive disadvantage in both US and Canadian markets, but also that Canada would become a less attractive opportunity for foreign investment."[79]

In Canada, competition over natural resources and their policy is disputed between provinces and the federal government. There is ambiguity and lack of clarity in the Constitutional Act of 1867 regarding this matter. Provinces have claimed exclusive jurisdictional rights over natural resources;[80] the federal government has claimed rights related to the environment, such as fisheries, sailing, federal lands, international treaties and topics of "order, peace and good government."[81] In addition, subsidiarity has been discussed in the courts. As

77 Bramley, 2000.

78 Harrison, 2010a.

79 Harrison, 2007:2–3.

80 Juillet and Patterson, 2009:106–7.

81 MacKay, 2004.

Harrison notes, Canadian courts have granted authority for environmental policy to provinces under the "property and civil rights" clause, granting more autonomy to them than to the federal government.[82]

Canadian federal governments have been very active in designing policies involving climate change. In 1990, there was the Green Plan; then in 1995 came the National Action Program; in 2000 they discussed the Action Plan 2000, followed by the Climate Change Plan for Canada of 2002. However, GHGs continued to increase.[83]

In 2006, the newly elected conservative government of Stephen Harper announced that Canada would not meet its KP goals. However, the topic of climate change was revived in parliamentary discussions, and two bills were drafted. The first one was Bill C-30, Canada's Clean Air and Climate Change Act, intended to help Canada meet its Kyoto targets.[84] This bill was passed by Parliament in 2007, and it became the Kyoto Protocol Implementation Act.[85] There was also Bill C-377, Climate Change Accountability Act, introduced in the House of Commons in 2006 by Jack Layton, then the leader of the New Democratic Party (NDP).

During the electoral campaign of 2008, it was clear that Prime Minister Harper's major opponent, Stéphane Dion, leader of the Liberal Party, had a very clear position regarding climate change. Harper was against including the topic on the agenda; the context of economic recession of the US and Canada gave him arguments to put the topic aside. On the other hand, Stéphane Dion pushed the issue forward, but with very bad timing. In the context of economic crisis, his main proposal was to create a new tax, one that would aid implementing the KP goals (a national carbon tax). The defeat of the Liberal candidate expressed the preferences of Canadian society in a context of economic crisis: citizens had put the topic aside. When they were asked about the most important problem for Canada, the ranking of climate change dropped dramatically from 25 percent in 2007 to 7 percent by the end of 2008.[86]

One year later, when the country started to recover (in 2009), Bill C-377 was reintroduced as Bill C-311, and again, after a long debate in the Senate and the House, had to be reintroduced as Bill C-224 in 2011. The goal of this bill was to establish specific targets to reduce GHGs up to the year 2020 and in the long run to reach an 80 percent reduction by 2050.[87] This set of legislative initiatives was accompanied by important amendments to the Energy Efficiency Act, directed at controlling energy consumption products.[88] Up to 2012, there were GHG

82 Harrison, 2007.
83 Simpson et al., 2007.
84 Parliament of Canada, 2007a.
85 Parliament of Canada, 2007b.
86 The Environics Institute, 2011.
87 Parliament of Canada, 2010, 2012.
88 Natural Resources Canada, 2009.

regulations for the transportation and coal-fired electricity sectors, and those for the oil and gas industry were expected to be enacted during 2013.[89]

Nonetheless, in spite of these efforts, Canada has had very little success. In the first place, public opinion polls reflect the scant attention of citizens to the topic (from 2008 to 2011 the percentage of people who thought that climate change was important to the Canadian political agenda fluctuated between 8 and 7 percent).[90] Second, provinces consistently have different opinions regarding a national comprehensive climate law, due to their different energy intensity. Some provinces, such as Alberta and Saskatchewan, are producers and consumers of tar sands; others, such as Quebec or British Columbia, fuel their economies with low-carbon hydropower; yet others, such as Ontario, use nuclear energy in general. Oil and gas companies also tend not to favor GHG national regulations.

The facts show a constant increase in Canada's absolute GHG emissions since 1990. According to the Canadian Inventory Report, up to 2011, 79 percent of Canadian emissions were of CO_2. Of the total GHG emissions, the majority came from transportation (24 percent), followed by 33 percent from the oil and gas industries, and 11 percent from electricity. Canada accounts for the highest percentage of GHG emissions per capita in the region, with 20.4 metric tons of GHGs per capita up to 2011. However, its emissions intensity has registered a constant decreasing rate of about 2 percent annually from 1996 to 2011. Over the same period, GHGs had increased in almost 20 percent when compared with 1990 levels.[91]

The US: Climate and Energy National Legislation

The US was one of the cornerstones of the climate-change international agenda during the 1990s. During the first meetings under the UNFCCC, President Bill Clinton and Vice President Al Gore led the way to establishing the bases for the KP, together with Norway, Australia, New Zealand, Iceland, Japan and Canada. All these countries favored the inclusion of mechanisms for technological transfer and finance for emerging economies or underdeveloped ones, such as carbon sinks, and carbon markets for international credits.[92]

During this time, important efforts were made, for example the US Global Change Research Program (USGRP) launched in 1997 by the federal government. This program reported directly to the President's Office, and it had the goal of developing research on climate conditions (temperature and precipitation), impact and vulnerability of the US. Its strategy was to divide the country into 16 geographic regions. At first, this research program had

89 Stikeman Elliot, 2013.
90 The Environics Institute, 2011.
91 Environment Canada, 2012.
92 Arnaud et al., 2009; Jaccard et al., 2007.

good results. It produced several scientific reports regarding vulnerability and adaptation strategies for each region.[93]

However, elections interfered with this trend. George W. Bush gained office, and, in 2001, the US Congress decided not to ratify the KP. This put climate governance as a whole in question. Since then, climate change is no longer a priority on the Republican agenda.[94] Nonetheless, programs such as the USGRP continued to work, counterbalancing the strategy of some lobbies, which suggested that climate change had no anthropogenic source and that it was not a threat to humanity.[95] In addition, weak party discipline prevented Congress members from pushing forward the climate-change agenda due to local electoral costs. The Senate was not interested in ratifying the KP either; senators generally tend to protect the competitiveness of US business. As Harrison explained, "while the oil industry may be staunchly supported by Republican senators from Texas and Alaska, the interests of coal are defended in the Senate by both Republicans from Wyoming and Democrats from West Virginia, while Democrats from Michigan look out for the automotive industry."[96] Thus, she concludes, both Republicans (who feared that the US would lose competitiveness) and the labor part of the Democratic Party (who feared job losses) blocked ratification.

It was not until 2008, with the election of Barack Obama, that the topic was placed back on the agenda. In spite of the economic crisis of that year, public opinion gave a green light to the topic, as reported by a Gallup survey, in which 65 percent of the Americans interviewed were very concerned about global warming.[97] Obama used the recovery framework to include mitigation strategies directed at consolidating energy transition. He enacted the Economic Recovery Bill in which he granted 76,530 million USD for renewable energy, energy efficiency, clean technology, and for the restructuration of the USGRP.[98] This strategy was aimed

93 United States Global Research Program, 1997.

94 It is important to note that attention was directed to the topic of security, due to the events of 9/11.

95 This situation arose especially with the oil and gas lobbies, the coal industry and the automobile manufacturers. Entities such as the American Petroleum Institute, the US Chamber of Commerce, the National Association of Manufactures and the Global Climate Coalition presented strong opposition to ratification based on the assumption that the science relating climate change to human development was unproven. For further details, see Harrison, 2010b; Simpson et al., 2007.

96 Harrison, 2010b:77.

97 The years 2007 and 2008 (campaign and election of President Obama) show the highest percentages of concern about global warming since the year 2000. This trend was certainly aided by Al Gore's presidential campaign and the national (and international) tours of his documentary *An Inconvenient Truth*, launched in 2006. This survey also shows that in 2008 only 58 percent of the interviewees mentioned that the primary cause of climate change was human activity. To look at the time series for these and other questions related to the topic, see Gallup USA, 2013.

98 Weiss and Bramley, 2009:1–2.

to achieve energy independence—a position reinforced in all of the president's State of the Union Addresses since 2010.[99] The strategy of dealing directly with the sources of GHGs (the energy sector) had a double goal. The first goal was to establish a path towards self-sufficiency through energy efficiency; the second was to deal with climate change to reduce GHGs. Up to 2011, the most important sources of GHGs in the US were electricity and the energy industry, followed by transportation and industrial processes; the majority of the GHGs emitted represent CO_2 (83.7 percent). Up to 2011, the US had 15.8 metric tons of GHG emissions per capita, a value that has decreased steadily since 1990. GHG intensity has decreased steadily from 1990 to 2011. The general GHG emissions up to 2011 have increased by 8.4 percent when compared with 1990 levels.[100]

In June 2013, President Obama reaffirmed his 2009 commitment to reducing GHG emissions by 17 percent below 2005 levels by the end of the decade. The plan included the first-ever limits on carbon emissions from new and existing power plants. These are the single biggest source of carbon pollution, accounting for one-third of the US's GHG emissions and 40 percent of its carbon output. This proposal faces opposition in Congress. However, it can be activated under executive powers (with no need to go through Congress). To this date (2013), the fate of this proposal is uncertain.[101]

To deal with this pollution, the US Congress has discussed several climate proposals. There are various comprehensive climate-change bills under discussion, some of which have suffered transformations and have become new proposals. During 2007 and 2008 several proposals were introduced. The list accounts for Bill S. 2080, the Climate Stewardship and Innovation Act of 2007 by Sen. Joseph Lieberman (I-CT), Bill H.R. 620, the Climate Stewardship Act by Rep. John Olver (D-MA), the Sanders–Boxer Global Warming Pollution Reduction Act, the Kerry–Snowe Global Warming Reduction Act, the Bingaman–Specter Low Carbon Economy Act, the Dingell–Boucher Discussion Draft, and the Doggett Climate MATTERS Act.[102]

These bills were followed by Bill H.R. 2454, American Clean Energy and Security Act (ACES), introduced in 2008 by Democrat Representatives Henry Waxman of California and Ed Markey of Massachusetts.[103] The goal of this bill

99 The White House, 2010, 2011, 2012.

100 Environmental Protection Agency—US, 2012.

101 Rincon, 2013.

102 For a detailed description of goals and strategies of each of these bills, as well as for critical details, see Arnaud et al., 2009; McKinstry et al., 2008.

103 This bill was promoted by Henry Waxman when he was appointed chair of the House of Energy and Commerce Committee shortly after the presidential election of 2008, in substitution for John Dingell, part of the General Motors foundational family. Kathryn Harrison offers a detailed discussion of this matter (2010b). To see details of this bill, see Richardson, 2012; Pew Center, 2009c; Consilience Energy Advisory Group Limited, 2009:198–203; Arnaud et al., 2009.

was to incorporate 25 percent of renewable energy into electricity production for 2025. It was thought that ACES would create new, clean jobs, save money on energy costs, enhance US energy independence and cut global warming impacts. However, there was the fear that these national goals would endanger the flexibility offered by existing carbon markets.[104] The Waxman–Markey bill was approved by the House in June 2009, but never passed the Senate. One of the main points of discussion was the cost that the change to a clean economy would incur for average American people.[105] The other argument against the ACES Act was the uncertainty of the ongoing carbon regional markets (WCI—Western Climate Initiative—and RGGI—Regional Greenhouse Gas Initiative). This law aimed at creating a national carbon market with a cap-and-trade system, leaving in place regional initiatives up to 2017.

Going beyond this bill, there is the Clean Energy Jobs and American Power Act (S. 1733), proposed by Democrat Senator John Kerry (Massachusetts) and Democrat Senator Barbara Boxer (California) and introduced into the Senate in 2009. This bill proposes more aggressive goals: 20 percent reduction below 2005 levels for 2020.[106] In the same vein are the Cantwell–Collins proposal, called Carbon Limits and Energy for America's Renewal Act, the Bingaman bill, called American Clean Energy Leadership Act, and the Kerry–Graham–Liebermann[107] bill on climate change and energy. Some of these are still in debate, others have stalled, but none of them has been passed by both chambers. These initiatives also faced the mid-term election of November 2010, where the main issues were the financial crisis and welfare-system reform, not climate change. However, in 2011, another bill tried to find its way through Congress with the same bad results: Bill H.R. 3242, called Save Our Climate Act, was not enacted. An interesting twist took place after the federal election of 2012, when President Obama was re-elected. At least in the discourse, the topic was put onto the national agenda again by the president when he announced a proposal for car makers to produce vehicles that would do 54.5 miles per gallon by 2025. Up to 2013, this target remains an illusion.

Apart from the political barriers to climate legislation, other obstacles that prevent Congress from passing these bills relate to LG initiatives for carbon markets. These usually have goals and commitments different from those proposed nationally. The LG approach also differs from that of a comprehensive national effort. For example, California has been leading carbon-markets know-how and

---

104 Wall Street Journal, 2009.

105 The Environmental Protection Agency (EPA) estimates that the cost for households would be 88–111 USD per year. Gas prices were estimated to be around 0.25 and 0.77 USD per gallon for 2030. For further information, see Pew Center, 2009b.

106 Federal Energy Regulatory Commission, 2010; Samuelsohn, 2010.

107 Since April 2010, Sen. Lindsay Graham (R-SC) stepped away from her efforts to develop a climate change bill, due to a change in the priorities of Congress towards immigration reform over the energy/climate change issue. See Danish et al., 2010.

technology on the East Coast, which differs from the situation of the Texas oil production industry or the North-East industrial corridor. The diversity of the local economies throughout the country prevents taking a national standard or strategy, as in the Canadian case.

Mexico: Climate and Energy National Legislation

The only Non-Annex country in North America is Mexico. This country has participated in the Country Studies of the UNFCCC since the 1990s and, to date, it is the only Non-Annex country that has issued national communications regarding the state of GHGs in the country.[108] The strategy of the federal government has been the creation of institutions with a transversal approach. In 2008, the Intersecreterial Commission for Climate Change was created by nine ministries of state, the most important being those in charge of environment, energy and income. This commission designed the National Strategy for Climate Change and its implementation program.[109]

The National Strategy has two pillars: mitigation and adaptation. Regarding mitigation, it identifies opportunities for GHG reduction by making a sector analysis. The other pillar is adaptation, which involves dealing with the vulnerability of society and limited institutional capacity, especially at local levels. However, as in the cases of Canada and the US, in Mexico, the trend is not promising. The majority of GHG emissions (65.9 percent) represent CO_2 and come mainly from energy industries (45.1 percent), followed by transportation (22.2 percent), agriculture (12.3 percent) and other industrial processes (8.2 percent). Mexican GHG per capita up to 2010 was 6.7 metric tons, and, contrary to the trend registered by the US and Canada, has increased steadily since 1990. Up to 2010, GHG emissions in the country account for a 33.4 percent increase compared with 1990 levels. However, Mexico had a steady decrease in GHG intensity of about 6.6 percent from 1990 levels.[110]

The implementation program foresaw that GHG reduction for 2008–12 would lead the country to a decarbonization process without compromising national expectations for growth and development. As a result of these policies, the country has committed to voluntary goals, and it has proposed aspirational goals to reduce 50 percent of GHGs by 2050 (with 2000 as the base-line year). To reach these goals, the country aimed to reduce emissions by 50 million tons per year starting in 2009. In spite of these initiatives, reduction in GHGs in absolute terms has not yet occurred. In fact, emissions are increasing as a consequence of

108 Mexico has issued five communications: in 1997, in 2001, in 2002, in 2010 and in 2012.

109 Secretaría de Medio Ambiente y Recursos Naturales, 2009a and 2009b.

110 Instituto Nacional de Ecología, 2012:31–2.

economic growth. According to the last GHG inventory, Mexico had increased emissions by 33.4 percent in 1990 with an annual growth rate of 1.5 percent.[111]

In this context, the majority of LGs have enacted their climate-change laws. For example, there is the Climate Action Plan of Veracruz, the well-integrated proposal of Mexico City, or the climate-change plan of Chiapas. Based on these laws, in 2010, a National Law for Climate Change was introduced as a bill by Deputy Alberto Cárdenas Jiménez (National Action Party) in the House of Representatives. This bill proposed the creation of a National System for Climate Change to link different institutions related to the topic, to regulate instruments of public policy, to inform, and to create citizen awareness.[112] It is important to note that this bill proposed the participation of civil society and epistemic communities, as well as LGs, especially for adaptation policies.[113] Some other proposals were made. For example, a bill introduced by Deputy Gabriela Medrano of the Green Party focused on the sanctions for not accomplishing GHG national goals.[114]

The debate ended in 2012 when the Cárdenas bill was approved by both chambers (with several adjustments and modifications) and enacted as the General Law on Climate Change. This comprehensive law includes the reorganization of environmental bureaucracy and agencies, as well as linking different areas of the federal government, especially the Secretary for Energy. For example, the topic of climate change was included in the remit of the National Ecology Institute (now called the National Ecology and Climate Change Institute). The law enforces the environmental justice institution (Procuraduría Federal de Protección al Ambiente—PROFEPA) by mandating the creation of a National Emissions Registry. It also activates instruments to support the national mitigation and adaptation goals, especially to continue on a growth path but with fewer emissions. In this regard, it links already existing environmental instruments to new mitigating policies and agencies under a National System of Climate Change administered by the Secretary of Environment (SEMARNAT). In addition, it sets the competences of the three orders of government (federal, state and municipal).[115]

One of the most important goals of this law is the creation of emissions trading programs. This implies that each state and municipality will need to design its GHG inventories and registries using a standard methodology. As this trading system will be administered by the federal government, states and municipalities will need to form a special contract with the federation in order to operate.[116]

Challenges to the implementation of this law are evident. The states and municipalities have limited implementation capacities and have been operating their own local climate action plans. Some of them (especially the northern states)

111 Instituto Nacional de Ecología, 2012.

112 Cárdenas, 2010.

113 Ibid.

114 Partido Verde Ecologista de Mexico, 2011.

115 Cámara de Senadores, 2011b.

116 Cámara de Senadores, 2011a.

have already designed their action plans and inventories with funding from local US governments and agencies (such as the EPA). This means that the whole scheme is based on different methodologies from those proposed by the Mexican federation (which are similar to those of the KP). In addition, LG initiatives in Mexico—in contrast to those of Canada and the US—reach constitutional limits when trying to include the energy sector in climate policies, due to state monopolies in electricity and oil/gas. This situation is starting to generate conflict and cases are being taken to the federal court.

The only windows of opportunity to link climate change to the energy sector are the Renewable Energy Law and the Law on Sustainable Use of Energy, enacted in 2008 as part of a comprehensive energy reform. In the words of Isabelle Rousseau, this reform package was thought to promote the energy transition and somehow include the sustainability ingredient.[117] These laws had the goal of reducing the impacts of production and consumption of energy, use of water and soils, and remedying damages to areas related to the energy sector. They were inscribed into the National Strategy for Energy Transition and Sustainability, a document that proposes three specific goals: (1) to maximize the use of natural gas; (2) to include 35 percent of renewables in the energy mix; and (3) to stabilize energy prices.

The three national policies have tried to link energy with climate change by creating transversal proposals. However, national legislatures have had a more restrictive approach to GHG management than the market approach proposed by LGs. The three countries' national proposals (either at the stage of bills or effectively enacted as laws) have attempted to be comprehensive and include all GHGs and different economic sectors. Nonetheless, they have failed to deliver GHG reductions and to "encourage states to play a substantial and constructive role in the newly developed federal regime."[118]

In Canada, national legislatures have being competing with local ones. Some national Canadian policies have relied on the approval of some provinces, yet the majority have not received such support. In the US, national policies have generally been based on local ones, and have used either a complementary or a competitive approach. For example, California's high climate-change standards have served as a basis for the EPA methodologies for inventories. However, the low standards of Texas have being competing with the high national standards proposed by the Congress bills. In Mexico, the basis for climate law has clearly been some of the local climate-change initiatives. However, the link with energy policies is almost impossible, due to the state monopoly of the sector. It is noteworthy that, from 1990 to the end of the first decade of the 2000s, the three countries have had very different results regarding GHG emissions. The US has increased by only 8.4 percent, Canada by 18 percent, and 33.5 percent has been registered for Mexico. It can thus be seen that the country outside the KP scheme and the most developed

117 Rousseau, 2010.

118 McKinstry et al., 2008.

economic nation worldwide has been doing better in managing GHGs than those under the climate governance of the UNFCCC.

The US "success" has had to do with energy efficiency and the leadership of LGs to push forward laws and policies for green development and decarbonization. However, as noted in this chapter, the inclusion of renewable energy sources into national and formal regional policies is not a priority in the energy transition of North America. In addition, interconnected infrastructure for these sources is far from developed, as it is the conventional fossil sector. This situation prevents the three countries from advancing in effective climate-change strategies.

What alternatives are there for climate governance, faced with the limited action of the state (especially in Canada and the US)? As will be detailed in the following chapters, energy efficiency, the inclusion of renewables and climate-change policies have been promoted at different levels of regionalization, led by diverse actors other than national governments or trinational regional schemes.

Chapter 5
From Local Governments to Transnational Green-economic Regions in North America

Introduction

The global approach, formal regionalism and national policies in North America have proven to be limited when addressing climate change. Results for GHG reduction worldwide are far from decreasing and neither the NAAEC nor the NAEWG has promoted successful strategies. At a national level, the North American countries, except for Mexico, have failed to formulate a national law that sets specific goals and designs national mitigation or adaptation strategies. Thus the weak contribution of the global and formal regional approaches to the provision of global public goods seems to be a very significant failure. As presented in earlier chapters, such failures have one common explanation: the approaches to address GHGs do not take into account the economic differences *within* nation-states. In general, both approaches leave aside informal and differentiated mechanisms to deal with the provision of these goods.

Up to this point, the premise "for global problems, global solutions" has been questioned. The KP has been robust in the sense that it survived its expiry date and won an extension to 2020, but it has been ineffective in achieving any GHG reductions whatsoever. Likewise, formal regionalism in North America has reached its limits in dealing with climate change. This chapter addresses an alternative approach to providing public global climate goods, by assuming that LGs are the starting point for constructing a new concept of region. Thus the first section of this chapter explores the individual strategies of North American LGs to address climate change. It starts by analyzing the relation of LGs with their national governments regarding environmental jurisdiction and policy design for climate change and energy. This first section goes on to analyze the individual mechanisms that LGs use to deal with climate change.

The second section proposes new ways of conceiving regionalization processes, which serve as a tool to address climate change in North America. The regional approach is assumed to be the correct one, but what is missing is a more flexible concept of "region," where relocation of authority falls to other actors, for example LGs. The second section also presents the essential elements in the creation of the concept of "transnational green-economic region" (TGER). This section defines how the institutional architecture of these regions is designed, as well as the norms and rules that bind them together. This transregional strategy has been one of the LGs' mechanisms to deal with climate change in North America.

Local Governments and Climate Initiatives

In spite of the integration of the North American energy sector and the creation of environmental trilateral institutions, the intended implementation of their recommendations still relies on the sovereignty of each country. The performance of the executive powers and legislatures of the three countries in North America has not delivered national climate–energy laws that address the problem in a comprehensive manner. Facing this scenario, LGs have taken action.

Local Governments Facing Global Governance and National Legislation

There are several strategies that LGs follow to address climate change and link it to energy efficiency. The first deals with the direct links between LGs and global climate governance through "*direct participation in global climate-change forums.*" LGs have found that individual participation is important, but collective presence influences decision-making more effectively. Associations have been the channel for this aim.

The participation of LGs in global climate conferences and meetings is not new. Arising from the inertia of the Montreal Protocol to protect the ozone layer in 1988, some LG-based multilateral institutions were created. For example, the International Council for Local Environmental Initiatives (ICLEI) was formed in 1989 by 35 local government leaders from Canada and the US. They joined together to discuss the depletion of the ozone layer and to coordinate LG responses to global environmental problems.[1] The creation of this initiative was led by the city of Irving, California, funded (at that time) by the United Nations Environmental Program. ICLEI has been present in different Conferences of Parties (COPs) in the UNFCCC governance framework. At the COP-15 in Copenhagen, LGs participated in the Local Government Climate Change Leadership Summit, organized by the Government of Denmark, in close cooperation with many global and regional LG associations and international organizations.[2] Their participation proved that local climate action was a consistent part of national climate strategies. Thus, when thinking about any future multilateral agreement substituting for the KP, "decisions or agreed efforts on limiting global warming and combating climate change must highlight the need for a partnership between national and local governments so as to ensure consistent planning and implementation at appropriate authority."[3]

A more intense participation was seen at the COP-16 in Cancun, Mexico. Some weeks before COP-16, a meeting of mayors and governors took place in Mexico City. The goal was to establish a common position for LGs (cities, provinces

1 This meeting included the participation and advice of some atmospheric scientists. See International Council for Local Environmental Initiatives, 2010d.

2 International Council for Local Environmental Initiatives, 2009.

3 Ibid.

and states), to be included in the final document of COP-16. Their position was established in the "Pact of Mexico City," with ten goals.[4] With the intention to present the Pact of Mexico City at the COP-16, ICLEI and the United Cities and Local Governments (UCLG) sponsored a delegation of LG representatives "to ensure the continuous and strong presence of LG's contribution to international negotiations during COP16, influencing directly the negotiations towards the recognition and the establishment of a clear and defined key role for local governments against climate change."[5]

However, during the COP meeting, "LGs expressed their concerns regarding the reduction of the negotiating text which discuss[ed] the future of the convention, and where local government references were included previously but not in the newly proposed text."[6] LGs pushed hard to be included in the final resolution of the Subsidiary Body, which noted that it recognized the important role and value of participation of all stakeholders (governmental and non-governmental) in the UNFCCC process. During the last days of the COP, Marcelo Ebrard Casaubón,[7] President of the World Mayors' Summit on Climate, presented the results of the prior reunion of city mayors.[8]

This was the first time that LGs had a voice and could introduce initiatives into the final COP documents. In a very strange approach, LGs were recognized as "non-governmental actors" by the member states of the UNFCCC process. Although this situation was not technically correct, it reinforced LGs' commitment. The final statement of ICLEI also highlighted the fact that LGs had a stronger voice and were recognized as stakeholders. In the words of the Executive Director of ICLEI-USA, Martin Chávez, "[w]hile modest agreements were achieved this year in Mexico, there is no doubt that local governments continue to lead the way on climate action."[9] In spite of these efforts, LGs still have no vote in the COPs' process.

In addition, LGs participate actively in the Governor's Global Climate Summit. Its objective is to expand its collaboration network "to stimulate

---

4 The goals of the Pact were to: (1) reduce GHGs voluntarily; (2) implement mitigation strategies to achieve their goals in transportation, waste and energy efficiency; (3) develop adaptation strategies, especially for the poor inhabitants of cities, who are the most vulnerable; (4) register, report and verify goals through the registry, called Carbonn (Bonn Center for the Report of Local Climatic Actions); (5) develop linkages to direct international funds; (6) establish a Secretariat to give feedback on performance; (7) promote participation of civil society; (8) cooperate with multilateral (UNFCCC) and national institutions; (9) promote alliances and cooperation among cities; (10) invite other cities' representatives to join the Pact. For further details, visit the site of the World Mayors Summit on Climate, 2010. See also Milenio, 2010.

5 International Council for Local Environmental Initiatives, 2010c.

6 International Council for Local Environmental Initiatives, 2010b.

7 Mayor of Mexico City from 2006 to 2012.

8 International Council for Local Environmental Initiatives, 2010a.

9 Sustainable City Network, 2010.

economic growth, create green jobs, and increase the use of clean energy while reducing greenhouse gas pollution."[10] LGs also participate in the Global Compact, a worldwide organization sponsored by the United Nations to gather together all sorts of actors. Some LGs (provincial and municipal), companies, NGOs and other actors participate in this global forum.

From Cancun to Durban and Qatar, LGs have been present. Some of the relevant topics they have put on the discussion table are the need to address the urban climate agenda and the disconnection between existing global climate policies and what most LGs deal with on a daily basis: poverty, lack of infrastructure and funding, and water scarcity. The presence of LGs has become normal at these meetings. However, this does not mean that they have the expected impact on national policies or global climate governance. Nonetheless, as long as the claims of LGs are specific, the creation of policy options and instruments is easier to fund and develop. Another example of the LGs' presence is the concrete proposals presented at COP-17 (Durban, 2011) and COP-18 (Qatar, 2012). At COP-17, the improved Harmonized Emissions Analysis Tool (HEAT+) was presented. This is software that helps LGs to plan mitigation actions and manage their GHG emissions.[11] At COP-18 in Doha, Qatar, 114 mayors from 28 countries presented the Durban Climate Change Adaptation Charter, recognizing that, on the one hand, the impacts of climate change are felt at local levels. On the other hand, LGs could be better equipped to take rapid action and prepare for and adapt to the impacts of climate change.[12]

The project called URBAN-LEDS dealt with low-carbon strategies for eight cities in four countries (Brazil, Indonesia, India and South Africa). This project is aided by the European Commission and ONU-HABITAT, and represents an example of the linking performed between LGs and global governance schemes. Likewise, the Climate Change City Registry (CCCR) and the organization Carbonn presented their second report of activities, highlighting that more than 200 cities (part of these associations) emitted 1.5 Gt of GHGs up to 2012.[13] Here, it is important to note that LGs' presence at global governance institutions is evolving from the plea to take their role seriously to the presentation of concrete results. However, this action is still not enough when we consider that the impact of a few LGs cannot be compared with the aggregate impacts of global GHG emissions.

The second strategy that LGs have developed is to establish "ad hoc relations with national governments or formal regional institutions." As explained in Chapter 1, the relations between formal and informal norms can be of different types. In general, LGs adopt ad hoc forms to relate to national governments,

10 Governor's Global Climate Summit, 2009.

11 International Council for Local Environmental Initiatives, 2011b.

12 This Charter was primarily directed to aid LGs from the "South," especially those in Africa. To read the document, see International Council for Local Environmental Initiatives, 2011a.

13 International Council for Local Environmental Initiatives, 2013a.

sometimes complementing their activities, cooperating between layers of governance, substituting national norms, designing competitive legislation or simply negotiating the adjustment of national legislation.[14] In general, the absence of serious national leadership regarding climate change has pushed LGs to develop their own strategies,[15] which is the case in North America.

LGs in this region have been working on mechanisms to address climate change by having a more efficient management of the energy sector. As Barry Rabe notes, almost every LG has tried to develop "home-grown" energy sources to foster long-term economic development and reduce GHGs. This has resulted in the development of various policy tools that link energy to climate-change management.

In the case of the US, these activities have historically created somewhat tense relations between LGs and the federal government. Some US states have even been defending in federal courts their right to enact local legislation regarding either climate change or energy. The most representative case is the State of California suing the Environmental Protection Agency (EPA) over setting emission standards on vehicles and on renewable-energy portfolios.[16] Cathy Milbourn gives other examples, such as when "several states, local governments and environmental organizations sued EPA over the agency's failure to update the pollution standards for fossil fuel power plants and petroleum refineries, two of the largest source categories of GHG pollution in the United States."[17]

Imbalances between LGs and national policies in the US go beyond courts to the policy field. For example, as Dukert notes, in the US,

> state public utility commissions may issue rate tariffs that are counterproductive, various state legislatures may decree a hodge-podge of definitions for renewable energy that deter needed imports of electricity generated by large Canadian dams, and low-cost power utilities in the Southeast or Northwest may resist the creation of interregional power networks that could improve reliability while lowering average prices … The result at the moment is that subnational policies (province by province and state by state) are actually prevailing, producing division rather than harmonization.[18]

In the case of Canada, facing the fact that the KP's targets were not realistic and would not be achieved, provinces have played a leading role on climate change and energy policies. British Columbia, Quebec, Manitoba and Ontario have been developing climate action plans and energy portfolios with very high standards;

---

14 For a discussion on intergovernmental arrangements in federations, refer to Brown, 2002; Elazar, 1987.

15 Brown and Sovacol, 2011; Harrison, 2010a and 2010b; Rabe, 2010; McKinstry et al., 2008.

16 BBC News, 2008.

17 Milbourn, 2010.

18 Dukert, 2007:155.

even Alberta has developed its own (taking the lowest standard possible). This is facilitated by the fact that natural resources fall under the provinces' jurisdiction.

Sometimes, federalism can discourage cooperation between LGs and the federal government. For example, some local policies tend to discriminate against Canada's large electricity sector (hydropower) to favor the smaller dam systems already located within a US neighboring state.[19] These situations exemplify the difficulty of drafting national climate strategies and how federalism can act as an obstacle to policy-building. However, not all LGs act equally. For example, some Canadian LGs (such as Quebec and British Columbia) participated in the establishment of goals for the country at the KP, but others, such as Alberta and Ontario, tried to block ratification to protect the oil and automotive industries.

In the case of Mexico, the relation with the federal government has been cooperative. The leading role of addressing the climate-change issue has also fallen to LGs. Some Mexican federated entities, such as Mexico City, Jalisco, Veracruz or the northern states, have developed their own climate action plans in cooperation with the BECC and NAAEC institutions; some others have designed theirs with the aid of Mexican federal agencies or funding by foreign ones. The federal government has supported these policies through the Ministry of Environment (SEMARNAT). The topic of energy has a different approach. In Mexico, energy is managed by two state monopolies: PEMEX (Petróleos Mexicanos) and CFE (Comisión Federal de Electricidad). The Mexican Constitution states that natural resources belong to "the nation" and should be administered by the federal government. This situation makes it very difficult to link climate-change local policies to federal energy ones, as explained in Chapter 4.

Facing these several challenges to intergovernmental cooperation, LGs have adopted alternative strategies to address energy efficiency and climate change in the region.

Local Governments' Individual and Regional Climate Governance

The third strategy used by LGs to deal with climate change is "regulation of climate and energy policies through several instruments." These instruments fall under climate or energy categories. There are climate instruments, in which climate action plans are cornerstones of regulation. The process starts with the development of GHG inventories and registries. In the case of North American LGs, they tend to choose and combine more than 250 instruments.[20] One of the most popular instruments is carbon caps, in which LGs establish regulations for the maximum allowable volume of CO_2 emissions (caps), either for a diversity of sectors (especially the energy sector) or for government facilities. California

19 Dukert, 2007:155.

20 McKinstry et al., 2008:4; Knigge and Bausch, 2006. For examples of each of the US states regarding energy efficiency and renewables policies, see the database of State Initiatives for Renewables and Efficiency, 2009.

has been the leader in this mechanism. Carbon taxing is another example of a regulatory instrument used by LGs, where a direct price for carbon is set by governments[21] based either on the carbon content of the fossil fuels (coal, oil, gas) or in terms of dollars per ton of CO_2e emitted. British Columbia is an example of carbon taxation, where, to be effective, the tax increases constantly.[22]

There are also "energy instruments" that deal with the mix of energy resources necessary to accomplish climate goals and efficiency. Among the most popular are the *Renewable Portfolio Standards (RPS)*. These require including a specific percentage or amount of electricity generated from eligible renewable or clean sources into the energy mix of the jurisdiction. In other words, they require setting GHG goals in time and achieving them by increasing the output of renewable energy from sources such as solar, wind and geothermal energy.[23]

Other mechanisms are related specifically to energy consumption, such as *public benefit funds*. These funds are allocated either through a small charge on the bill of every customer or through specified contributions from utilities.[24] Some of these include funding for renewable development and/or for efficiency goals (such as that of Montana or New Mexico); others are focused on carbon capture and storage (CCS), subsidies for efficiency of appliances, low-income household energy assistance, or investment in renewables (as in the cases of Quebec, British Columbia and Oregon). There are also *energy efficiency standards for appliances* not covered by federal laws, such as in the case of Washington (13 more appliances than the federal law) or Rhode Island (19 more appliances).[25] Energy consumption instruments often promote market development. Examples are *green pricing* and *net metering systems*, as in the cases of Arizona, Manitoba and New York. These mechanisms transit from consumption to production when they allow customers to sell electricity back to the grid (net metering) and to pay a special fee to electric companies to assure that a certain percentage of the electricity they receive comes from renewables (green pricing).[26]

Likewise, there are mechanisms dealing with energy consumption in the transportation sector. For example, some states and provinces, such as Illinois, Ontario or South Dakota, use *financial incentives for biofuels and alternative fuels* to implement lower-carbon performance of the fuels used in transportation. In other words, the gasoline sold must contain a certain percentage of renewables. Also, some others, such as Maine, Maryland and Manitoba, use *vehicle GHG and renewable standards* that require vehicles to comply with certain GHG goals.[27]

21 Meckling and Hepburn, 2013.

22 Aldy and Stavins, 2011; Duff, 2008.

23 Pew Center on Global Climate Change, 2009a; Knigge and Bausch, 2006; Zachary and Carpentier, 2003.

24 Pew Center on Global Climate Change, 2009a:3–7.

25 Knigge and Bausch, 2006.

26 Pew Center on Global Climate Change, 2009a:3–7.

27 Ibid.

The fourth strategy has been a regionalization process and the creation of regional associations. There are different examples, such as the Conference of New England Governors and Eastern Canadian Premiers, the 20 Region Club, the Network of Regional Governments for Sustainable Development, or the Cities Climate Leadership Group, among others. Each of these organizations is led by an LG. However, some of these processes go beyond and create hybrid (formal–informal) regions with market mechanisms to link the economic activities and the natural resources available with climate–energy policies through the integration of green-economic transregional schemes. In North America, there are examples, such as the Regional Greenhouse Gas Accord, and the Western Climate Initiative, where LGs are the main policy actors. In order to construct the TGERs' institutional architecture, LGs must harmonize some of their instruments to manage GHGs. This guarantees the possibility of LGs joining associations or linking to federal legislations with a low cost. These transregional experiences have become innovating building blocks to address energy and climate policies and test the robustness of re-territorializing the provision of global public goods.

Traditionally, the regional approach has been thought of as a level of analysis, as explained in Chapter 1. A region conceived as a dynamic space and constructed with a transnational focus would envisage institutional alternatives to the provision of global public goods. In this way, a region is no longer a level of analysis, but a layer in which there is interaction with formal and informal institutions and actors. This means that the location of a leading authority needs to be elsewhere: neither in international institutions nor on national levels, but on transregional layers. Acting like building blocks, transnational regions in North America seem to be experiencing a more fruitful path than the other approaches. This chapter explores theoretically how a transnational approach can merge with regional governance to provide global public goods such as clean air in North America.

To address this question, let us think of a scenario where LGs, provincial or state governments integrate transnational governance schemes and become the fundamental actors in reconfiguring climate governance and its strategies to reduce GHGs in North America. The process of transnational integration led by LGs results in new regional alliances based on the harmonization of their energy and climate-change policies. This integration is led by the economic particularities and the trends in transregional integration (especially within NAFTA). This chapter proposes that the entities leading the reconfiguration of North American climate policy are the transnational green-economic regions (TGERs).

Reconfiguring Regions and Regionalization

Regions can be understood as "spatial expressions of ideas useful for the analysis of the problem at hand … [which is] to separate into recognizable component parts the otherwise overwhelming diversity and complexity of the earth's surface."[28] In other words, a region is conceived as a framework for organizing and ordering common characteristics within a certain space. The region could then be understood as a "territory where a system of governance exists."[29]

Within any region there are two attributes that are interrelated: the physical characteristics and the human factors. The physical characteristics include the natural attributes of the location and all natural resources within a system; they are usually called "ecoregions" or "natural regions."[30] These regional conceptions of space are based on different types of natural resources. Regions are defined under climate criteria, under biodiversity or under landforms based on geological features.

There is a deep connection between ecoregions and human population. The former sets the conditions for humans to exist and develop, while the latter modifies them in doing so. Every human activity leaves its imprint on the landscape and natural resources through interconnections and interrelations that change constantly.[31] Connectivity influences the spatial diffusion of ideas, products or services that generally converge in the relation between nature and humans within the region. Although no place on earth is equal to another in physical terms, interconnections of people and places show patterns of spatial similarity, allowing us to identify common elements.

Since a region can be conceived as a social construct, there can be many regions over the same territory. This is the reason for establishing some of the general characteristics that every region has: (1) location; (2) spatial extent; (3) boundaries; (4) hierarchy.[32]

Location can be absolute or relative. Absolute location expresses the exact position of a point on a map, using degrees of latitude and longitude. Relative location refers to the location of a place in relation to that of other places or things. In other words, a region is a territorial entity that can group political entities, but that can also be independent of political borders.[33] Relative location would then reflect spatial interconnection and interdependence in dynamic ways, whereas spatial extent may vary. Spatial extents are defined by the size and the scale of the region in relative terms.[34] This characteristic leads to the establishment of boundaries within which common governance criteria can be established. Different regions overlap the same territories, so they need to be hierarchically arranged based on the main criteria and their basic common pool of variables.

There are different ways to view a region. Based on the common-pool variables, one can find economic criteria and probably think of regions as commercially

28 Getis et al., 1991:415.

29 Catenacci, 2010.

30 Bailey, 1998:2; Hettne, 2000:157.

31 Getis et al., 1991:7.

32 Ibid.:416.

33 Stevenson, 1989.

34 Getis et al., 1991:4–5.

integrated blocks of nation-states, such as NAFTA; others can focus on military power and form regions as a result of security agreements, such as the North Atlantic Treaty Organization; some others can understand regions as links between people and migratory phenomena as region connectedness.[35]

In environmental studies, it is common to have different sets of common-pool variables. For instance, ecoregions address the natural links between flora, temperature, humidity and so on. However, if one is interested in species conservation, the starting point would be a particular species—or even a group within a species. The boundaries of the region would be set in relation to the species' needs and movement. If the species migrates in search of food or water, the region's boundaries have to follow this direction. Another example is the integrated basin approach to water management. In this case, the region would be delimited by the way a river flows from the top of the mountains to the ocean or lake. The last example of ecoregions is natural protected areas. In this case, the boundaries are set in administrative terms, setting limits according to the social, economic and political status of the area.

As noted above, an ecoregion can then set boundaries in either static spaces, such as the natural protected areas, or in dynamic and moving spaces, as in the case of basin water management or species conservation. Spatial extent is not always linked to territoriality. For example, we can find virtual regions where technology creates the "space" for interconnection or establishes administrative management, such as regulations on space or international oceans.[36] In this case, the traditional concept of territory has been challenged by globalization. In economic terms, globalization has broken the localization–production link, leaving the power of decision-making to capital, not to territorial governments. In cultural aspects, globalization has also threatened territorial traditions and created hybrid communities. Politically, institutions (formal or informal) can become weak when facing global trends, especially those social movements that fight for non-territorial causes, such as human rights or gender. These processes interfere with the traditional functional operation of the territory.[37]

These four factors also represent a simultaneous encounter between continuities and discontinuities. Discontinuities appear with the constant change of societies. This change can have different time frameworks. For example, change in historical ages, where concepts of society, government and economic wellbeing are driven by grand ideas, moves to other sets of images and beliefs. Discontinuities can also be less broad, as in the case of power relations during the nineteenth century and the Congress of Vienna. On a more specific level are those changes of territoriality expressed as sets of norms and rules, as in the case of climate governance. In general, discontinuities tend to change the concept of territory and the compounds (regionalization initiatives) that are formed

35 For a detailed definition of "region states" see Ohmae, 1995.

36 Ferreira, 2005.

37 Held and McGrew, 2002.

around it. Discontinuities can also be understood as flexibility to accommodate differences in institutional schemes.[38] On the other hand, continuities can be present when we find established links that tend to be resilient to change. Some examples are common agreements, integration of economic activities, people's day-to-day exchange of goods, services, values and perceptions, and power relations. Continuities tend to find it difficult to accommodate changes.

However, we must not assume at this point that institutions are obstacles to change or that they exemplify the status quo. They favor change in different forms (reform, defection and reinterpretation), as explained in Chapter 1. In fact, processes of institutionalization create and destroy regionalization processes all the time. In general, resilience of a regional architecture is possible thanks to the malleability and interaction of formal and informal institutions. It is fair to acknowledge that the element of discontinuity tends to push forward resilience in a homeostatic format, and continuity when institutional architecture resembles homeorhesis.[39] In both cases, a process of change has occurred, but it is not clear when a transformation process will develop into a new governance scheme that will change the concept of territoriality.

Based on this argument, regions then become territories in constant construction, with articulations (dynamic or static, flexible or rigid) of ecosystems, and social interests, goals and practices with different levels of authority, and, for formalized ones, legal constraints. The region can be understood as a spatial organization framework. In other words, a region can be a "unity not in the sense of static content, but in the manner of its operational connectivity."[40] The complexity of these connections can be explained by different elements that interrelate in the process of construction and operation.

Two elements for the definition of regions must be addressed: the structure and rules under which they perform. In other words, it seems necessary to analyze the institutional architecture that gives structure to regions and the understandings under which they are assembled.[41]

---

38 Hofman and Mérand, 2012.

39 Homeostasis is the natural tendency of a living organism to maintain a stable equilibrium, in the thermodynamic sense, among its interacting internal components, regardless of interactions with the external environment. For a dynamical system, homeostasis can be viewed as sustaining internal stability. In contrast, homeorhesis is the natural tendency of a living organism to continue its evolving development, albeit varying under different environmental conditions. For a dynamical system, homeorhesis is viewed as maintenance of order and normalcy in the state trajectory under internal and external disturbances by avoiding a potential chaos. These definitions are taken from Ray and Phoha, 2005.

40 Getis et al., 1991:416.

41 For a detailed discussion about institutional architecture, see Hofman and Mérand, 2012.

It is therefore important "to separate the players from the rules."[42] The players within the regional architecture are very diverse. They go from governmental to non-governmental actors and come in a variety of institutional shapes and colors. In addition, a region can also become an acting subject with a distinct identity, actor capability, legitimacy and a set of rules for decision-making.[43] There are different ways to address this distinction. As mentioned in Chapter 1, some scholars have analyzed the formal and informal characteristics of structures and rules in regions. Recalling prior definitions and applying them to a regional approach, an example of formality would be to develop an agreement among national governments driving the regional integration processes; these agreements link state actors through the enforcement of rules.[44] The formation of a region of this type tends to be identified with a static and rigid political project (coming from the state) and is usually called "regionalism."[45] Informality would be conceived as a process of regionalization that can act through other channels, especially direct and private interactions among societies and non-state actors. Thus regionalization involves a variety of state and non-state actors, who often come together in rather informal multi-actor coalitions,[46] emulating dynamism and flexibility.

The links and connections these players perform are the norms under which the regional structure operates. Rules can also be categorized as formal or informal. As explained in Chapter 1, norms are the instrument by which to delineate behavior. Regional norms then construct the region itself. A "normative" region generally includes a territorial jurisdiction over which social interactions are performed. What is interesting about the possibility of conceiving regions as social constructs is the fact that norms link the four elements of a region (location, spatial extent, boundaries and hierarchy). In addition, norms define the level of dynamism or statism of the region's structure. The "creation" of a normative region leads to establishing the policy domains for interaction. In the case of climate change, the European Union is an example. As mentioned in Chapter 2, some countries in Europe signed up to the KP as regional actors. This example assumes that actors in the structure can perfectly separate the different regions in which they participate in policy domains. However, this is very difficult to assess. Overlapping regionalization processes, with different rhythms and shapes, complicate independent decision-making. In addition, regions have the characteristic of being dynamic and static simultaneously, which makes it almost

42 North, 1990:4–5.

43 Hettne, 2000:157–8; Rosenau, 1997.

44 Boesen, 2006:2; Boas et al., 2005:1; Knight and Calvert, 1995.

45 The differences between regionalization and regionalism are explained in Söderbaum and Shaw, 2004:7–8. Bjorn Hettne also analyzes this distinction by proposing that there are levels or steps towards "regionness," defined as trans-local sameness that gives coherence and regional identity to the regional project. Hettne, 2004:28–9.

46 Söderbaum and Shaw, 2004:1–2.

impossible for the actors to benefit from the overlaps. Nonetheless, regional norms sketch the framework of regional governance and help overcome this problem.

In the case of climate change, regionalization processes tend to be diverse. Some of them, as explained in Chapter 4, represent formal structures that form regional agreements with several institutions, such as the NAAEC. This regionalization process was led by national governments through official channels. In other words, it was designed hierarchically from "above." Informal regionalization schemes from "below" contested the commercial focus of NAFTA, pushing forward this institution. The immediate question to address deals with other "directions" for creating a region. Put another way, who can design regions? How should regionalization processes in climate change be categorized when they do not come from "above" or from "below"? Is it useful to think of middle-range directions, especially when dealing with the provision of global public goods?

To try to respond to these inquiries, this book argues that regional governance from the "middle-and-around" can be a useful approach to dealing with climate change. This middle ground includes agreements signed by LGs, such as provincial ones, city mayors, municipal or community leaders, as well as links among social sectors of the economy. This middle-range approach tries to accommodate elements of formality and informality within the construction of transnational regions in North America regarding the policy domain of climate change and energy. The compound "climate change and energy" is per se a strategic and a practical issue. As a strategic issue, it is of public interest and manifests itself in enforced agreements. As a practical issue, it requires infrastructure supported by formal agreements, but it also performs connectivity between societies and sectors that transcend national borders and establish informal links in transnational ways. As Roberta Capello and others suggest, "a local treatment of sustainable development falls in between a macro (or global) systems level and a micro (or project) level approach, as the analysis of local or urban sustainability is clearly a meso level analysis."[47]

Transnationalism, in this sense, will deal with diverse local actors (especially LGs) trespassing national boundaries to meet their supranational goals. Transnationalism does not imply the disappearance of the state, but emphasizes that states are only one type among other actors. This concept suggests the need to explore a "re-governing" world where territorial grounds and national autonomy or sovereignty cannot be taken for granted.[48] It has become evident that the governance scheme in NAFTA is far from being supranational, as it is in the European Union. However, other regions "outside the box" fulfill some of the criteria for European supranationalism. Transnational regions formed around climate change and led by local governments have the capacity to "construct mechanisms that would generate a self-sustaining dynamic at the continental level," mentioned by Clarkson.[49]

47 Capello et al., 1999:9.

48 Djelic and Sahlin-Anderson, 2006; Mann, 2000:140–42.

49 Clarkson, 2010:290.

Institutionalization of these mechanisms through transgovernmental compounds helped to construct transregional governance. As with the Commission of the European Union, transregional governance in North America implies institutions in which decision-making power has been ceded by governmental authorities, yet from a different level of government. Thus LGs are creating supranational institutions without the consent of the state as the grantor of ceded sovereignty.[50]

In other words, this regionalization process refers to the trans-local relations of a social, political, cultural and economic nature between human compounds.[51] Sovereignty reappears as a perforated phenomenon in which a series of different actors, such as transnational companies, political parties, non-governmental organizations, territorial communities or individuals, respond to trans-sovereign signals. Authority is also perceived as diffuse among a variety of actors and issues, either global or local. As Isidro Morales points out, "state authority is being overlapped, in its own territory, and beyond it, by multi-leveled layers of authoritative institutions that remain well differentiated in spite of being intertwined with the state."[52]

Although transnational linkages present a constant and dynamic flow of ideas, goods and services that are difficult to materialize, there are other levels of authority and actors that constitute the basis for these dynamic exchanges. This was one of the main problems in making the KP work: it was assumed that global responses would need to territorialize in nation-states. This kind of territoriality—based on the concept of classical sovereignty—proved a failure. For this reason, we need to ask how we can re-territorialize responses to deal with global public goods in climate change. This process means the redefinition of territory and sovereignty, and the relocation of authority for climate-change governance.

This section has conceived the region as a construct for organizing common criteria. A region is constituted by structures and rules that can be local, national, transnational or global, and that can interact in different directions. A region can set dynamic or static boundaries, depending on the starting point. With different starting points at hand, it is understandable that we can find that different regions overlap. The focus of this research is to define regions with specific criteria for addressing climate change by establishing the causal link with energy, and analyzing LGs' strategies. This means setting hierarchies for environmental problems and assuming that the study of climate change is a priority because of the unavoidable link with energy policy. As the UNFCCC suggests, climate

---

50 These governance arrangements must not be confused with what Anne-Marie Slaughter and Thomas N. Hale call "transgovernmental networks." These networks account for official informal connections linking governments with other actors abroad. As soon as these links are transformed into formal agreements, these relations are no longer considered as networks. For a detailed discussion of transgovernmental networks, see Slaughter and Hale, 2010.

51 Hettne, 2004:27–9.

52 Morales, 2008:20.

change is linked with and influences almost all other environmental problems, due to its range and scope. The uncertainty of its intensive impacts is another reason why it needs to take top priority. Yet the transnational response of local actors to climate change is a phenomenon not fully explored in the literature of international relations: it appears as an alternative route with practical opportunities to address the problem of global GHG reduction through a regional approach.

The following section explores how transnational regions have been constructed and operated in North America within the climate-change policy domain based on local action. In other words, it explains the relations of LGs in a transnational governance framework. In this regard, it helps explain how transnational regionalization of climate change in North America is being run by "transnational green-economic regions" (TGERs). These kinds of regions can be considered as regions emerging within regions,[53] with their own governance mechanisms, and their own sets of common-pool variables, rules and actors.

Institutional Architecture: Structure of TGERs

In North America, there are two different kinds of regions. The most evident kind is multinational, represented by NAFTA and explained above. Less evident are TGERs. Harald Baldersheim and Morten Ogard argue that "the emergence of new policy options for regional development are seen as creating pressures on existing institutional structures, driving a search for new organizational patterns better suited to the pursuit of the new policies [where] new policies require new partners."[54] This statement explains that, facing the lack of national comprehensive legislation to deal with climate change in effective ways, other actors fill this policy gap. In general, private actors (especially business and NGOs) have taken on this task or have been assigned to it by the same governments. However, TGERs fill the gap with hybrid institutions (not merely part of national policies, not merely part of private initiatives), expressed in their own particular architecture.

The structure of TGERs is fourfold. The first element of the structure is the composition of this type of region. Although regions can be defined in non-geographic terms with little emphasis on location,[55] it is my aim here to explore a more conventional approach linking regions to territory, and to explain how the territorialization of global climate governance performs in transregional North America. In addition, I will explain how local governments of provinces and states become the main units of TGERs through the creation of institutions that address climate change. LGs can be seen as the main actors of the re-territorialization of policy responses and of the relocation of authority to address this problem.

---

53 Hettne, 2000:157.

54 Baldersheim and Ogard, 2011:3.

55 Mansfield and Milner, 1999:591.

As regions are "always in the making,"[56] they tend to include a variety of actors that participate in this process. LGs have shown the capacity to adapt their governing machinery to build relations with different actors at different levels and layers that respond to global climate pressures, as well as for designing and implementing policies to cut GHGs in the short and medium terms. However, an institutional design for the reconfiguration of climate governance, based on a local–regional approach, would have to build strong relations between institutions, society and business. The main goal of this form of governance would be the co-production with their citizens of collective global goods.[57] LGs can direct other actors to implement global ideas at a local level. In the case of climate change in North America, this assertion implies that national governments or other governance schemes have failed or fallen short in managing GHG reduction, as presented in chapters 2 and 3.

Brian Hocking suggests that we have to go beyond the perspective of exclusivity, discontinuity and separatism in the relations of LGs[58] with other actors. As he explains,

> [w]e are confronted by an ever widening "frontier" where domestic and foreign policy meshes into a seamless web and in which differing "spheres of authority" jostle one another for advantage, then non-central governments [LGs] can, by virtue of their actorness, be seen as patrolling the edges of this territorially imagined and yet partially de-territorialized space.[59]

The fact that LGs can be considered as actors means that, conceptually, they can be the starting point for constructing the analysis. In addition, it implies the need to create innovative solutions and models for future decision-making and legislation within a participative framework.[60] LGs have created spheres of authority that have defied, complemented or competed with national policies and other schemes of governance. Thus "new patterns of governance may entail pressures on the legitimacy of public authorities, requiring new transformative strategies to enhance citizen understanding of public action."[61] LGs tend to represent the last window of public representation, where demands can be pushed forward or contested democratically. LGs become the closest public authority a citizen has. TGERs link

56 Boas et al., 2005:1.

57 Hart and Pakrash, 1999:15–16.

58 This author uses the concept of *non-central governments* to refer to what I call local governments. He explains that this is the most accurate way to name the units we are studying. He argues that this term does not imply hierarchy, but that the national government is a centralized one, which also offers conceptual problems for my case studies.

59 Hocking, 1999:17.

60 McKinstry et al., 2008; Capello et al., 1999.

61 Baldersheim and Ogard, 2011:5.

demands not only on a jurisdictional basis, but also on a regional basis, by trying to stay as close as possible to the citizens.

Another factor that determines the composition of the region is the economic link among LGs' jurisdictions and citizens. TGERs have been born in a context in which transnational economic linkages have performed in North America even before NAFTA (yet deepened by it). Indeed, "regions in Canada are economically integrated not with some amorphous mass called the United States of America, but, rather, with particular American regions and states."[62] Relations between transborder societies have historically existed in US–Canada and US–Mexico borderlines, performing vertical links, as exemplified in previous chapters by energy links and trade in North America.[63] As Clarkson notes, these linkages develop because individual states' or provinces' shared geography makes them vulnerable to the behavior of their neighbors, especially in relation to environmental issues.[64] This fact has influenced the type of economic activities in the region and changed intra-regional commerce. Productive relations have an important role in the constitution of TGERs because they set specific environmental standards and integrate sectors such as industry, energy or transport. Jurisdictions smaller than the nation-state, but linked regionally, are also responsible for environmental impacts.

Environmental impacts and externalities generally are evident in neighboring jurisdictions. Therefore the second constitutive element of TGERs' structure deals with how close the main actors are in physical terms. TGERs can establish links between neighboring units within a country or across borders. In addition, these regions can have transnational linkages when joining neighboring units in a wider area. The regional approach helps to share the responsibility of environmental impacts, linked to economic integration in a larger territory. Thus

> [l]arge (i.e., territorially extensive) [regions] have the virtue of exploiting economies of scale in the provision of public goods, internalizing policy externalities, allowing for more efficient taxation, facilitating more efficient redistribution, and enlarging the territorial scope of security and market exchange.[65]

Physical proximity also implies that, in practical terms, infrastructure to provide public goods is easier to connect in order to share the costs and benefits of their production. For example, in the case of the energy sector in North America, almost all fossil fuel sources tend to have dense networks that connect the whole area. Having a TGER implies that some parts of these networks start to be transformed into more sustainable ones. However, proximity and the capacity to interconnect infrastructure can play in the other direction.

---

62 Van Nijnatten and Boychuk, 2006:487.

63 Ramos, 2002; Hiernaux, 1993; Gunderson, 2001.

64 Clarkson, 2008:134.

65 Hooghe and Marks, 2003.

This argument leads to the third element of the structure, "interconnection." Interconnection means that there is energy infrastructure that connects physically different areas and LGs, either for consumption or for production. Energy infrastructure includes (1) power plants for electricity generation and transmission systems, (2) pipeline systems for natural gas, gas storage and liquefied natural gas facilities, and (3) refineries, pipelines and import facilities for oil.[66] In addition, energy infrastructure accounts for renewables exchange facilities, such as dams or transmission lines from wind turbines.

At this point it is important to clarify that energy interconnectedness deals with both consumption and production. Consumption and production can have direct or indirect links with climate change and GHG emissions. Consumption is generally present in transportation, industry and electricity. However, energy consumption of electricity is linked to climate change only indirectly. Consumers, in general, do not know what part of the electricity they use comes from fossil fuels or from renewables. Turning on a lamp, per se, does not emit CO_2. However, it becomes a direct source of CO_2 when we understand that, for example in the electricity sector, it does not matter if consumption is low or high once electricity has been produced. If a whole city turns off the lights during one hour, the supply of electricity will keep running and will be wasted. There is no technology that can store electricity to balance supply and demand. In this case, production of electricity appears as a direct source of CO_2 emissions when produced by fossil fuel sources.

Either way, consumption and production of electricity function on the basis of interconnectedness. The North American Electric Reliability Corporation (NERC) is an example of transregional interconnectedness established in North America.[67] It divides the territory into several electrical areas, as shown in Figure 5.1. This division guarantees the efficiency of the service based on reliability standards, which "are the planning and operating rules that electric utilities follow to ensure the most reliable system possible."[68]

Energy consumption also includes direct burning of fuels, especially for transportation, industrial processes or production of electricity. In this case, fuel consumption is a double GHG emitter: it is a source of CO_2 when burning directly and an indirect source when used to produce electricity. To understand North American TGERs, it is fundamental to know the kind of energy connections

66 Prado, 2007:406.

67 The NERC was established in 2006. The members of the regional entities come from all segments of the electricity industry: investor-owned utilities; federal power agencies; rural electric cooperatives; state, municipal and provincial utilities; independent power producers; power marketers; and end-use customers. These entities account for virtually all the electricity supplied in the US, Canada, and a portion of Baja California, Mexico. The US states now belong to different reliability councils, the Midwestern Reliability Organization, the Reliability First Corporation, Southwest Power Pool, the Southeast Reliability Corporation, and the Northeast Power Coordinating Council for the Canadian provinces.

68 North American Electric Reliability Corporation, 2011a.

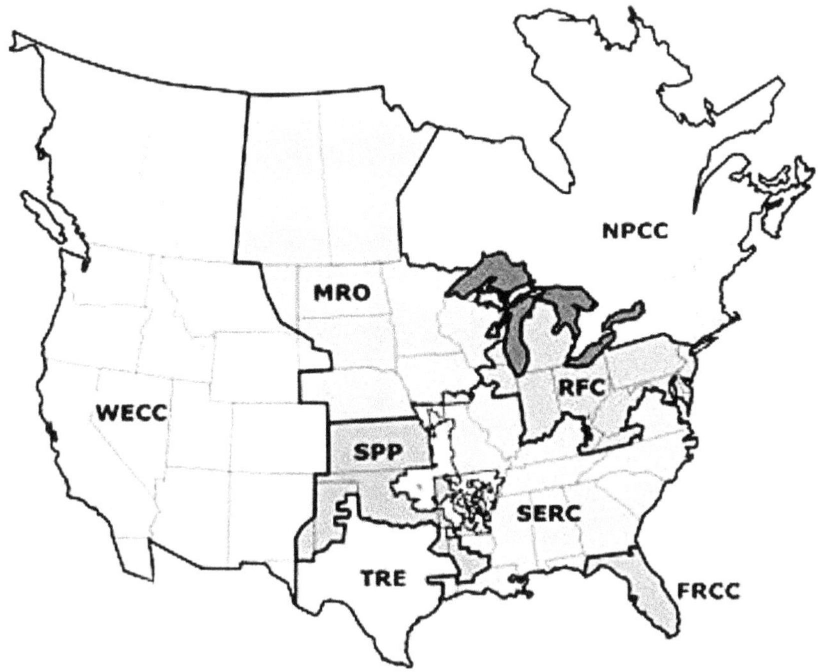

Figure 5.1 NERC regions
Source: North American Electric Reliability Corporation, 2010.

LGs have. This will contribute to the delimitation of TGERs' action based on the regional distribution networks. An example of this is the hydro capacity in the Pacific Northwest, which generates large amounts of electricity in excess of the region's need (and, therefore, large transfers to other regions) when river flows are typically highest in spring and early summer.[69] Chapter 6 presents the details of TGERs' connecting capacity.

The third element that completes the structure of TGERs is the "geographical and climate diversity" in North America. The structure of the region will also depend on the kind of resources present.[70] For example, for energy production it is necessary to map mineral resources such as oil, coal, gas, uranium and other kinds of chemicals used in the process. For climate-change mitigation policies, the presence of vegetation, such as forests, is required to absorb CO_2. However, the uneven distribution of these resources, due to technology and transportation costs,

69 Energy Information Administration—US, 2013.
70 Conca, 2005.

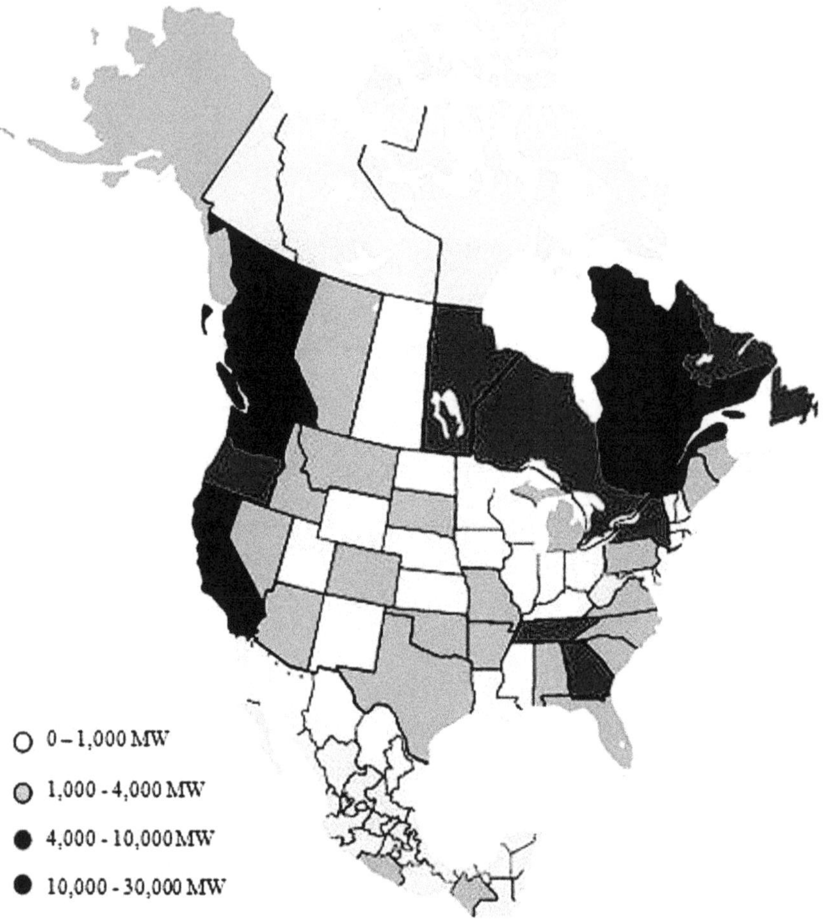

Figure 5.2 Renewable energy resources installed in North America
Source: Adapted from Commission for Environmental Cooperation, 2003.

influences the energy production of the region. Examples in Figure 5.2 show the situation regarding different renewable capacities built within the region.

If mineral resources are scarce or inaccessible, TGERs need to develop other strategies to acquire them, such as importing them or integrating their energy networks with others. As the Energy Information Agency notes, low-cost electricity generated with nuclear and coal-fired capacity from the Commonwealth Edison service area flows over transmission lines through the Midwest region and back into the eastern portion of PJM, a local regional transmission organization in the District of Columbia. This intraregional flow amounted to about 100 million MWh in 2010.[71]

In order to be consistent with an environmental approach, TGERs need to find mechanisms to avoid externalities derived from imports of this type. If a region is rich in hydro capacity for energy production, means for energy distribution and consumption must be in place, so the region's dependence on local resources will be high. Another example is if a forest needs to be preserved, the cooperation of related organizations is necessary. In this case, "areal relationships to industrial concentrations and the impacts of material extraction upon alternate uses of areas are important to the definition of the region."[72] Víctor Toledo and Patricia Moguel propose that, in order to achieve optimization and better use of resources within the natural context, it is necessary to understand the type of ecogeographic unit in which they are situated, as well as their productive potential, and the desire for their preservation.[73] In addition, it is necessary to analyze the environmental criteria of the territory in order to preserve it and manage GHGs.[74]

In North America there are 15 types of ecosystems, ranging from the cold tundra in the north, the forest mountains in the Rockies, the Great Plains, the deserts, to the tropical forests. These ecosystems are shaped in vertical forms, similarly to the exchange of energy.

In spite of its rich diversity, North America has very few areas with a high degree of environmental protection. The major protected areas are located in a corridor on the West Coast, from Alaska all the way down to Northern Mexico. In general, conservation is located in forest and desert lands.[75]

The structure of TGERs involves LGs integrating regions based on physical proximity aided by infrastructure interconnectedness. In general, TGERs tend to match with certain ecosystems. The main premise of this chapter is quite simple: GHG emissions are the product of energy resources used by economic activities. Thus regional institutional design is needed to give coherence to any climate policy.

Institutional Architecture: Norms of TGERs

The regionalization process of TGERs needs to have a set of norms and rules in order to operate. These norms express common objectives, detail the mechanisms of performance, and establish sanctions. In other words, "[l]inkages

71 Energy Information Administration—US, 2013.

72 Getis et al,. 1991:430.

73 Toledo and Moguel, 1992.

74 Smith et al., 2004:9; Capello et al., 1999.

75 There are diverse forms of conservation policies for natural areas. There are national parks, natural reserves, habitat management areas, protected land/seascape and management resource areas, among others. For a detailed map of the main natural protected areas in North America, see the North American Atlas at Commission for Environmental Cooperation of North America, 2012.

and the distribution of attributions and capabilities are mainly done through the codification and enactment or rules, treaties and organizations, in which the degree of compliance [can be] flexible and variable."[76]

In the case of climate change in North America, norms have several integrative characteristics. The first one is the fundamental idea of "sharing climate-change goals and strategies." These regions recognize that they can provide global public goods using a transregional approach. However, they also acknowledge that their economies and energy needs are not equal. Because of this, they have established flexible mechanisms for members to draw up their own goals. Schelling warns us about the importance of not losing focus. These goals (general and particular) should be directed to the problem of GHG reduction and management, not to success in the integration process.[77] Therefore climate-change regional governance is a tool for a bigger aim, but in order to become operational, it must draft clear mechanisms, as specific goals for GHG reduction or measurement models. Transregional governance also must create clear instruments and methods to achieve the GHG reduction goals. For example, they can opt for cap-and-trade systems, carbon taxes, energy portfolios or green markets. In this way, they share the idea of including environmental scope in their regional policies. The common feature of the TGERs of North America is the will to create carbon markets through the cap-and-trade system; this represents coherence with the vision of climate policies in the US, where the market approach prevails, *vis-à-vis* the European model, where there is a more restrictive focus.[78]

The sharing of knowledge and information is also basic for the setting of goals and the functioning of strategies.[79] Epistemic communities or transnational informational groups that share expertise generally base their performance on others' best practices, "emulating the policies of those that are competitors for foreign direct investment or even international portfolio capital; and harmonizing policies with trading partners so that there is as little 'friction' or conflict over policies as possible."[80]

In the case of North American TGERs, these shared ideas, goals and strategies do not yet imply identity-building. What emerges is a model for coping with consumption and production rules. Nevertheless, "the design of a successful micro-constitution can create community where community did not previously exist."[81] It is only with the long-term sharing of common environmental values in relation to energy saving and efficiency that an "environmental" way of life can emerge in these regions.

76 Morales, 2008:21.
77 Schelling, 1993:481.
78 Antal, 2004; Vargas and Valdés, 2007.
79 Conca, 2005.
80 Stubbs and Eaton, 2006:373.
81 Ostrom, 1992:348.

The second characteristic is the "intensity of integration of policies and mechanisms," especially those linked to energy consumption and production sectors, such as industrial processes and transportation.[82] This integration implies specialized rule design. In fact, regulation and convergence policies can be the result of direct integration in order to facilitate exchange, but sometimes they represent the indirect by-product of other aspects of integration.[83] Integration can be uni- or multisectoral, general or specific. Authority, then, relocates differently depending on what path the integration takes. As Cerny notes, "particular economic sectors which have developed autonomous self-regulatory governance processes will become the predominant authoritative allocators of collective goods."[84] Within integration of policies, it is also important to incorporate external environmental impacts of other actors (externalities) into the regional policy schemes to absorb costs and make policy adjustments. In this sense, these regions integrate energy policies with climate goals in a transnational way, a situation that results in the common sharing of externality costs. In addition, sector integration is generally transnational in the sense that TGERs relate with global problems and with different levels of authority and actors. Their scope of action sometimes trespasses national borders, fragmenting sovereignty, but it sometimes stays in the local sphere. TGERS do not ignore national or global arrangements; they try to integrate them into their unity when they find useful elements to work with.

In North America, there are areas that have very high levels of industrial pollution. Some of these can clearly be grouped as geographical regions, whether within the borders of territorial provinces or states (like Alberta) or representing a group of them (such as the Atlantic North-East Coast). According to the North American Environmental Atlas database designed by the CEC, the majority of industrial activities or energy production/consumption in the region matches with the CO_2 flux.[85] In other words, pollution clusters coincide generally with energy production and consumption and that required for industry performance. As presented in previous chapters, in the three countries of North America, the energy sector accounts for the majority of GHG emissions.

Were we to overlay in a map the fossil fuel resources, industrial pollution areas, renewable zones and the ecoregions, it would be evident that areas with no

82 Rowlands, 2008; Bernstein et al., 2008.

83 Gunderson, 2001.

84 Cerny, 1999:207.

85 The North American Atlas database can be consulted at Commission for Environmental Cooperation, 2012. According to data from the GHG National Report Inventory of Canada, tar sands in Alberta account for the majority of CO_2 emissions. However, in the majority of CO_2 flux maps, it does not appear as a major emitter. Total CO_2 flux is estimated as the sum of two CO_2 sources (fossil fuel emissions and vegetation fires) and two CO_2 sinks (terrestrial ecosystems and oceans). Because of this kind of measurement, Alberta does not appear to have a high carbon flux. Alberta's emitting sources are the tar sands. For a detailed explanation on how CO_2 is measured in Alberta, see Environment Canada, 2008b.

rigorous environmental protection are generally different from those places where pollution clusters appear. For example, on the Atlantic East Coast pollution cluster there are no relevant environmental restrictions.

The third characteristic for region-building is the creation of "institutions for operating policy choices for integration." For the case of climate change in North America, TGERs have designed new transnational institutions. In this discussion it is pertinent to reintroduce the dichotomy of formal–informal. Institutions in TGERs are formally constituted through agreements among LGs. However, these agreements are based on certain informal characteristics, such as reliance on local information, the possibility to be effective and implemented in small-scale economies, and high entry barriers. For LGs to integrate a TGER, they must include institutions that can design GHG inventories and climate policies within their jurisdictions. In addition, they need to align local regulation with the new standards. These technical conditions set the bar high for some of the LGs in which climate is not a policy priority.

To manage the cost of creating the know-how to operate this institutional architecture, TGERs rely on other transregional initiatives, those not necessarily focusing on the environmental area. Borrowing institutional models needs to address carefully the nature of the region, and pick those elements that are coherent with the transnational focus. This is why only a few instruments from global climate governance are useful for TGERs.

Nonetheless, TGERs need to create instruments to prevent the costs of uncertainty when other orders of government or governance global schemes come up with a new proposal. TGERs need to be flexible in the event of institutional change. This is why the final element is "linkage with other governance schemes," especially in the case of new national legislation and a new binding agreement when the KP expires in 2017. Thus transregional initiatives need to have the capacity to act as flexible building blocks in a "middle-range" approach. TGERs must be willing and prepared to perform as bases of larger carbon markets and link with other actors of global climate governance.

In general, a TGER reinvents the borders and scope of failing global climate governance. This must involve a process of re-territorialization and relocation of authority to obtain a different approach to climate governance. The basis of this process takes into account a different concept of region in North America, one articulated on a transnational basis. The TGERs' structure integrates different factors, such as location and spatial extent. However, the borders try to link political jurisdictions with natural-resource areas and economic activities. This link allows LGs to integrate with others, especially their neighbors. Structural factors also deal with the capacity to interconnect energy distribution mechanisms, basic for economic integration. These regions encounter a binding instrument represented by norms and agreements. In general, the main goal of the North American TGERs is to manage climate change and energy policies in a more cost-effective manner than other proposals.

As explained in this chapter, in North America there are tight economic links among LGs, where regional experiences start to serve as a basis for integrating important parts of their economies and their energy–climate policies. Nonetheless, integration varies from region to region, as well as according to environmental impacts and GHG emissions. Jurisdictions manage climate policies in a variety of ways. Some produce high levels of GHG emissions because of their dependence on fossil fuels for producing energy; others tend to reduce fossil fuel consumption by developing clean energy options. The economic diversity and difference in emissions of GHGs have driven LGs to use diverse instruments to address climate policies. Among them, they have created TGERs with the previously described characteristics. Chapter 6 explores how LGs managed to integrate this kind of region in three cases in North America.

Chapter 6
Transnational Green-economic Regions: Two Carbon Markets in North America

Introduction

As mentioned in previous chapters, one of the strategies of local governments (LGs) to address climate change has been to enter into regionalization processes through the integration of transnational green-economic regions (TGERs). In general, these processes are institutionally hybrid. They are informal in the sense that they are not led by national governments or the state, and they transcend borders, linking political jurisdictions elsewhere. They are formal because they work under institutional architectures where they "cede their sovereignty" to an administrative entity with enforcement capacity. They are not subordinated to either national governments or global governance institutions, but design mechanisms of linkage. In other words, they create their own transregional governance mechanisms.

This chapter focuses on two cases that better exemplify the creation of TGERs, the Regional Greenhouse Gas Initiative (RGGI) and the Western Climate Initiative (WCI). These TGERs formed carbon markets to give incentives to the transition to sustainable economies and energy efficiency while not sacrificing their local development. These initiatives, although newly created, have shown institutional capacity to perform.

The main instrument LGs use to integrate TGERs has been the creation of carbon markets and the use of the cap-and-trade mechanism.[1] To clarify the technicalities of these types of markets and their instruments, the first section of this chapter explains how carbon markets work, as well as several mechanisms such as the cap-and-trade system, common to RGGI and WCI. The second section presents RGGI and WCI separately, their origins, the integration of policies, the

1 Some authors include carbon markets in a wider category called "renewable energy markets," working on a wider scale of polluters, policy options and links with other policy areas such as development, education or trade. Here, the main difference is that the carbon markets increase the use of renewable energy to reduce GHG emissions and the goals and regulation are generally established by governments. The renewable markets work under renewable credits on the basis of exchange of a wider scale of polluters (e.g. particulate matter or mercury) and with no emission targets. As they recognize, there is still no clear definition of the differences between these, which leads to constant overlap. For further details, see National Renewable Energy Laboratory, 2006.

results of the cap-and-trade mechanism, the relation with other governance actors and the integration of infrastructure.

The recurring example for both cases is the electricity sector. This sector is one of the most highly emitting ones in North American economies, so it is a good example of the performance of climate and energy policies in each of the two cases. Projections for the North American region reveal that there will be an increase of consumption up to 2035.[2] The two transregional initiatives, being aware of this fact, have designed mechanisms to incorporate cleaner sources of energy in the electricity portfolio to move towards energy efficiency framed under the proposed energy transition.

The last section of this chapter deals with several critiques to this transregional approach, which reveal some of the weaknesses of re-territorializing the provision of global public goods.

Regional Carbon Markets and Cap-and-trade

As explained in Chapter 5, LGs have a variety of options to manage GHGs, ranging from regulation (of technology choice, emissions, consumption, or production of energy etc.) to pricing the externalities through carbon markets. There are some common policy instruments to price carbon and create carbon markets. For example, there are environmental standards, carbon taxes, emission reduction credits, clean energy standards, fossil fuel subsidy reduction, renewable energy credits and emissions trading schemes (ETS).[3] Within this last category fall carbon markets in North America.

Types and Features of Carbon Markets

The main goal of a carbon market is to set a price for carbon (or CO_2e) and give incentives to businesses to reduce and manage their emissions.[4] This goal is achieved by commoditizing GHGs and putting a price of exchange. There are voluntary markets, where no law enforcement is established and the relation among market participants is totally private and informal. In general, these markets are based on voluntary carbon credits. Some examples are when airlines offer to neutralize customers' carbon footprint or when companies choose to obtain

2 Energy Information Administration—US, 2010a.

3 Aldy and Stavins, 2011; National Renewable Energy Laboratory, 2006; Capello et al., 1999.

4 These types of markets are called "carbon" markets, although they include a variety of GHGs. In general, "carbon equivalent" or "CO_2e" refers to the equivalence of other GHGs in CO_2 units.

certificates to be seen as socially responsible organizations. The main reasons for this are reputation and social benefits.[5]

There are other markets that are mandatory in the sense that they involve governmental enforcement and generally hold formal governance relations. There are several conditions in mandatory carbon markets that policy-makers must determine. The first deals with the number of allowances a company can have and the type of GHG sources to regulate, either upstream (based on carbon content of fuels) or downstream (based on monitored emissions).[6] In some carbon markets, the number of allowances given to the companies has no limit. This is a double-edged situation. On the one hand, it appears to provide incentives to business to enter into the environmental dynamic; yet the price of the allowance in reality is transferred to the final consumer. In other words, in unlimited-allowance markets, the consumer usually pays the price of business GHG emissions. On the other hand, when allowances are limited, companies are encouraged to innovate and assume the cost of technological change.

The second condition is whether or not these allowances are sold in auctions or provided free. Selling allowances guarantees that at least a specific percentage of the profit goes to public funds for energy efficiency, environmental conservation, GHG reduction, or research and development of clean technology. For this to be efficient, the government needs to collect revenue identical to that from a tax producing the same amount of emission abatement. A looser cap translates into setting a lower tax and vice versa.[7] If allowances are provided free, companies will only be doing very profitable business without being committed to manage or reduce GHGs. As noted by Aldy and Stavins, to give free allocation of allowances could reflect some historical record or "grandfathering" involving transfers of wealth, equal to the value of the allowances, to existing firms, whereas, with an auction, this same wealth goes to governments.[8]

Figure 6.1 presents existing carbon markets and the type of activity they perform. Japan, New Zealand, Australia, Norway and the UK operate their markets under the guidelines of the KP; the European Union ETS does too, but started operations early in 2005, before the KP was fully ratified. The US regional markets operate outside the KP trading scheme. In general, all markets start with a free-allowance approach and turn into a sold allowance when the market has settled down.[9]

The case studies in this research are categorized into the most controlled scheme, limited number of allowances and sold at auctions. RGGI and WCI are compounds of LGs' commitments, which have developed several individual mechanisms that try to build up into transregional policies and instruments. Some

5 For details on these types of market, see Seeberg-Elverfeldt, 2010.

6 Aldy and Stavins, 2011.

7 Meckling and Hepburn, 2013; Aldy and Stavins, 2011; New York State, 2010.

8 Aldy and Stavins, 2011.

9 For a comparison of worldwide carbon markets, see Aldy and Stavins, 2011; Consilience Energy Advisory Group Limited, 2009; Arnaud et al., 2009; Flachsland et al., 2008.

	LIMITED # Allowances	
FREE Allowances	Japan, Swizerland, Chicago	European Union* Norway, Australia, RGGI, WCI
	United Kingdom (Phase I) New Zeland	United Kingdom (Phase II)
	UNLIMITED # Allowances	**SOLD Allowances**

Figure 6.1 Types of carbon markets

Source: Author's elaboration based on Lee, 2011; Aldy and Stavins, 2011; Ellis and Tirpak, 2006:10; and CRC Magazine, 2009 (for the case of the UK).

* In the first two phases of the EU ETS, emissions were distributed freely (Phase I, 2006–2008 and II, 2008–2012). For Phase III (2013–2020), auctioning of a larger share of allowances of the member states is required. In the international emission market the benchmark price is set by the European Union ETS, except for RGGI and WCI.

of these instruments are instituted by LGs, and others by markets in the forms of third-party incentives. Among the first are performance standards, renewable portfolio standards, taxes, public funds and cap-and-trade. Third-party incentives can be achieved through labeling schemes, green pricing, net metering and so on.[10] TGERs generally set the basis on government-instituted instruments, and complement with the others.

A third condition is the control of cost uncertainty, when allowance prices are volatile or high. There are containment measures that try to encourage political support for carbon markets. In this case, cost uncertainty can be addressed by offsetting excessive emissions with credits from emission reduction instruments lying outside the cap. Uncertainty can also be managed by banking-and-borrowing schemes, where firms can save an allowance for future use or bring a future-period allowance forward for current use. There are also safety valves to address uncertainty, such as putting ceilings on prices or establishing price collars (ceiling and floor prices). In addition, to address the loss of competitiveness between a firm under any of these instruments and another one that is not, the cap-and-trade system allows a "border adjustment" in the form of an import allowance requirement. In this way, imports can be working under the same regulatory costs as domestic allowances.[11]

10 For details of each of these mechanisms, see Zachary and Carpentier, 2003.

11 Aldy and Stavins, 2011.

In general, for a carbon market to be competitive requires the existence of a large number of potential traders where there are sufficient differences in abatement costs among such traders to make it worthwhile.[12] In other words, to create incentives for firms and jurisdictions to participate in a carbon market, it needs to be profitable at some point. As explained by Jonas Meckling and Cameron Hepburn, the debate between pricing carbon by taxes or by markets has been in existence since the 1980s. Both instruments have interacted simultaneously; however, according to these authors, the market approach has prevailed in the twenty-first century. One explanation for the market preference is that coalitions are easily put to work, but it is very difficult to organize a constituency to support any kind of tax. Environmental policy choices, in democratic societies, generally are determined by their perceived distributional effects and costs of implementation.[13] As explained in Chapter 4, the electoral campaign of Stephen Dion in Canada based on a carbon tax found no society support whatsoever in a context of economic and financial crisis.

The Cap-and-trade Mechanism

According to Aldy and Stavins, a cap-and-trade system constrains the aggregate emissions of regulated sources by creating a limited number of tradable emission allowances, in sum equal to an overall cap, requiring those sources to surrender allowances to cover their emissions.[14] When cap-and-trade programs are put to work, LGs utilize auctions to allocate allowances. In this scheme, TGERs use a multi-sector approach that includes electricity, industry, transportation, and residential and commercial fuel use; they also take into account forestry and agriculture. Let us not forget that TGERs operate in a format of sold and limited allocation of allowances, where

> [local] governments place a cap, or limit, on the total amount of emissions allowable, and then require firms to account for their emissions by obtaining emission "allowances" for every tonne of carbon dioxide they emit. Firms can obtain the allowances when the allowances are first allocated and later by buying and/or trading allowances with other firms.[15]

Cap-and-trade systems represent an indirect way to price carbon where the price is established by the market (contrary to carbon taxes, where the price is set by government). In general, the cap for emitting GHGs is controlled by either

---

12 Capello et al., 1999.

13 Meckling and Hepburn give a detailed explanation of how environmental policy choices distribute wealth and minimize costs. However, these are different by industrial sector and region (2013:479).

14 Aldy and Stavins, 2011.

15 Ministry of the Environment—Ontario, 2009:23.

governments or governance institutions. For example, in RGGI and WCI, the cap is either managed by the general council of each TGER, or established by member LGs. The logic of these carbon markets is that, in the long term, allowances (or polluting permits) given to the companies are expected to be more expensive and scarce. Pricing GHGs intends to limit energy production (fossil fuels) and consumption,[16] giving incentives towards energy efficiency, technological transformation and environmental conservation.

The cap-and-trade system favors compensation instruments among transregional members, emulating the model of the European Union in the KP. TGERs utilize offsets and secondary markets. Offsets allow entities to compensate their emissions with credits from emission reduction measures outside the coverage of the cap-and-trade scheme.

This instrument was originally designed as a means of allowing economic growth in non-attainment areas while insuring no further degradation of their air quality.[17] Arnaud et al. note that the first company to use offsetting was the US power company AES Corporation. In 1989 this company decided to finance an agro-forestry project in Guatemala by investing 2 million USD for planting 50 million trees to offset emissions of a power plant that it built in Connecticut.[18]

To operate offsets, first an offset provider needs to estimate a polluter's emissions. Calculations of offsets vary from one provider to another. In other words, they vary depending on the sector, region, project or methodology used. In addition, the buyers vary, too. They can be companies, non-profit organizations, governments and so on.[19]

A different approach to offsets has also been used in parallel. The "bubble" approach encourages intra-firm offsetting or offsetting within a certain group of sources. As explained by Capello et al., the offset allows transfers from existing sources to new ones, as the bubble policy allows transfers only among existing emitters (in general related among each other).[20]

The offset system encourages the creation of secondary markets. These markets provide spaces to exchange allowances abroad. Andrew Hedges notes that, for a secondary market to be a successful source for offsetting, it should be comprehensive, create scarcity and have predictable prices, as well as have clear regulation and enforcement rulings. Comprehensiveness means having several jurisdictions and sectors participating to avoid market manipulation, leakage to unregulated markets and uncertain carbon pricing. It also has to be able to create scarcity to encourage investment in emission reductions within installations and the development of new low-carbon installations. That objective is undermined

16 This ocurs only if there are some incentives for consumers to pay high prices for GHG-emitting energy sources.

17 Aldy and Stavins, 2011.

18 Arnaud et al., 2009:274–5.

19 For details on calculation of offsets, see Arnaud et al., 2009.

20 Capello et al., 1999:30.

where there is insufficient scarcity of allowances to underpin a strong price for such investments over a sufficient time period to justify them. In this sense, a secondary market, ideally, has to have predictable prices in the long run.[21]

Case Studies: RGGI and WCI

This section exemplifies the creation of TGERs and the operation of their carbon markets. Although they have common ideas about GHG reduction, their institutional architecture is not that similar. In addition, their scope and linking mechanisms also differ. RGGI, for example, to date has not fully reached a transnational scope, integrating only US states and having Canadian provinces as observers. WCI has gone beyond linking jurisdictions on the West Coast, from Alaska to California, with some Mexican states as observers. Thus WCI is transnational in the sense that it includes LGs from different neighboring countries. Although RGGI does not have this characteristic, it can be considered as a transnational region because it links to climate-change mechanisms beyond the state.

Each of the regional initiatives is analyzed according to three variables. The first two are recalled from Chapter 5. They account for the institutional architecture of each TGER regarding the institution-building process—the strategies, goals and instruments utilized to address climate change. The third variable opens up the discussion to be presented in Chapter 7. It explores the ways in which TGERs create linking mechanisms with other governance layers. This characteristic explains how flexible these initiatives are and how well they address the uncertainty peculiar to carbon markets.

The Regional Greenhouse Gas Initiative: An Operating Market

The Regional Greenhouse Gas Initiative (RGGI) is a transregional corporation that works in the region of the North Atlantic coast of North America (see Figure 6.2). RGGI started operations in December 2005 through the signature of a Memorandum of Understanding among seven US states (Connecticut, Delaware, Maine, New Hampshire, New Jersey, New York and Vermont). Two years later, Massachusetts, Rhode Island and Maryland were incorporated.[22] Up to February 2011, RGGI had several observers, such as Pennsylvania, Washington,

---

21 Hedges, 2009.

22 The Memorandum of Understanding was signed by Governor M. Jodi Rell (R), Connecticut; Governor Ruth Ann Minner (D), Delaware; Governor John Elias Baldacci (D), Maine; Governor Martin O'Malley (D), Maryland; Governor Deval Patrick (D), Massachusetts; Governor John Lynch (D), New Hampshire; Governor Richard J. Codey (D), New Jersey; Governor George E. Pataki (R), New York; Governor Donald Carcieri (R), Rhode Island; and Governor James H. Douglas (R), Vermont. In June 2011, the House and Senate of New Hampshire passed a law for withdrawing from the RGGI. However,

DC, Ontario, New Brunswick and Quebec.[23] It is worth noting that the Governor of New York, where the headquarters are located, led the creation of this initiative. This regional program was "built on state policies enacted earlier in Massachusetts and New Hampshire, and it was launched via strong leadership from New York and former Governor George Pataki in making the case for a regional collaboration."[24]

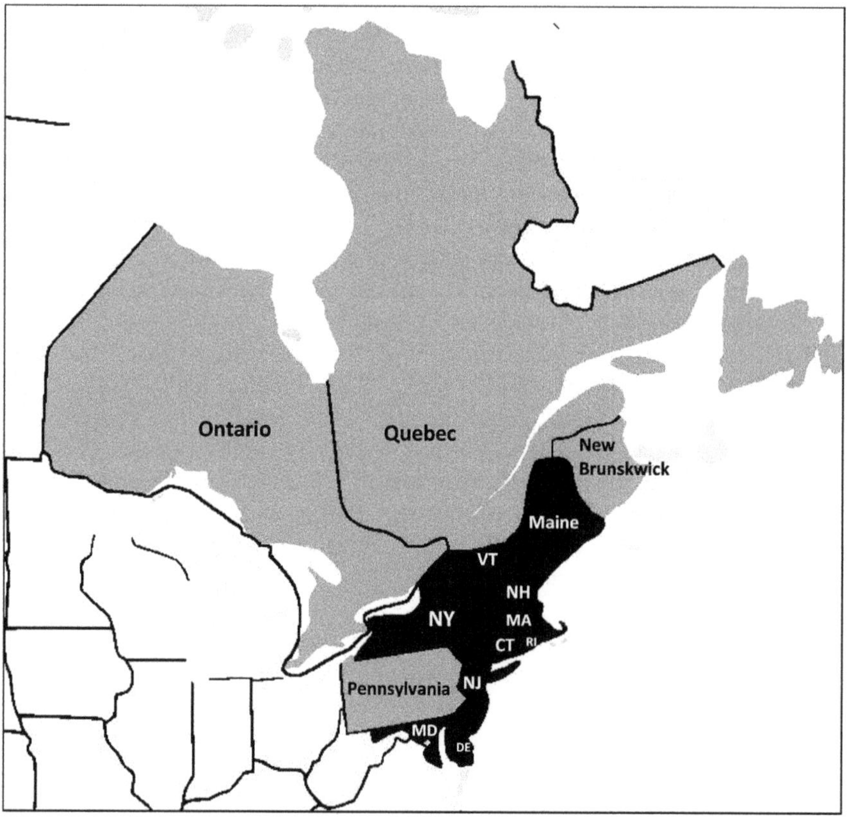

Figure 6.2 Geographic location of RGGI members
Source: Adapted from Regional Greenhouse Gas Initiative, 2010b, 2011b and 2013a.
Note: New Jersey participated only up to 2011.

Governor John Lynch (D) has said that he will veto the bill. For further details, see Danish et al., 2011b.

23 Regional Greenhouse Gas Initiative, 2011b.

24 Rabe, 2010:4.

RGGI represents 10 percent of the CO_2 emitted in the US. It also represents 16 percent of the population and 24 percent of GDP in the US.[25]

The carbon market in practice: Policies, strategies and instruments

The logic of RGGI integration is to add up individual members' efforts to integrate a compound. In this way, individual mechanisms reinforce the transregional strategy of RGGI. Table 6.1 depicts the set of instruments and emissions reduction targets each member of RGGI develops individually.

The main goal of RGGI is to develop a carbon market based on the cap-and-trade scheme for CO_2 emissions coming from the power sector.[26] The goal was divided into two phases. The first one would be to stabilize CO_2 emissions for 2014 at 2009 levels; the second would be to reduce them by 10 percent by 2018 (2.5 percent per year) in 200 electric plants that work with fossil fuels and that generate more than 25 MW. This calendar started on January 1, 2009.

Different actors participated in the process of developing RGGI. Some of them were local energy and environmental agencies, experts from the International Standards Organization (ISO), research institutes, business corporations and civil society. New rules and institutions have been created in this process. In March 2006, the "Model Rule" of the institution was presented. It was put for discussion by public opinion, participants and more than 100 non-governmental organizations. The results were some modifications and the draft of the Memorandum of Understanding to be enforced that same year.[27] The Model Rule has been adjusted on several occasions. The latest one took place on February 7, 2013, as a result of the revision stage one year earlier. The results of this revision were favorable, but they claim the need to incorporate new mechanisms to deal with low allowance prices and open up new offset possibilities for compliance.[28]

RGGI operates under a cap-and-trade scheme.[29] To facilitate its implementation, RGGI created "CO_2 Budget Trading Programs" (CO_2-BTP), in which it established

25 See World Resources Institute, 2010; Western Climate Initiative 2010a; and Pew Center on Global Climate Change, 2009a. See also the database of the Bureau of Economic Activities of the Department of Commerce—US, 2009, and Bureau of the Census—US, 2009. See also the Inventory of US GHGs at Environmental Protection Agency—US, 2008b.

26 This initiative does not cover the other five GHGs. Federal Energy Regulatory Commission, 2010.

27 Farnsworth, 2010:3–5.

28 One of the most important elements included in the 2013 Model Rule was a new requirement for sources to hold allowances to cover 50 percent of emissions for each interim control period. In other words, jurisdictions need to cover half of their emissions during the first year of the three-year period in each phase. In addition, any excess ton would be considered a violation of the agreement subject to enforcement by the LG responsible agency. For further details, see the Updated Model Rule at Regional Greenhouse Gas Initiative, 2013b.

29 RGGI uses "short tons" where 1 short ton = 907.18 kg.

Table 6.1 Type of individual mechanisms of members of RGGI

	Climate action		Energy sector			Transportation sector	
	Climate action plans (individual emission targets)	**GHG inventories and registries**	**Renewable portfolio standards**	**Public benefit funds**	**Green pricing and net metering (electricity)**	**Financial incentive for biofuel**	**Vehicle GHG and renewable standards**
CT	10% below 1990 levels by 2020	X	27% by 2020	Renewables and efficiency	X	X	X
DE	None	X	20% by 2019	Renewables and efficiency	X	None	None
ME	10% below 1990 levels by 2020	X	30% by 2000	Efficiency	X	X	X
MD	None	X	20% by 2022	Renewables and efficiency	X	X	X
MA	25% below 1990 levels by 2020	X	15% by 2020	Renewables and efficiency	X	X	X
NH	10% below 1990 levels by 2020	X	23.8% by 2025	Efficiency	Net metering	None	None
NJ	1990 levels by 2020	X	22.5% by 2021	Renewables and efficiency	X	None	X
NY	10% below 1990 levels by 2020	X	25% by 2013	Renewables and efficiency	X	X	X
RI	10% below 1990 levels by 2020	X	16% by 2020	Renewables and efficiency	X	X	X
VT	25% below 1990 levels by 2012	X	Equal to load growth 2005–2012	Renewables and efficiency	X	None	X

Source: Author's elaboration with data from Knigge and Bausch, 2006; Western Climate Initiative, 2010a; Pew Center on Global Climate Change, 2009a; 2010b. See also Government of Massachusetts, 2010.

that 25 percent of revenues would be destined for public benefit funds towards energy efficiency strategies or to control prices of energy. The remaining 75 percent would be used as the LGs decide is best suitable.[30]

There is fear that, with carbon markets, the price of electricity would increase or that companies would not be able to purchase allowances in time when they become increasingly expensive. According to RGGI's estimates, the increase in the electricity price will be of only 1.6 percent monthly (78¢) for residential use for 2015, and 2.4 percent up to 2021, taking into account the increase in the allowance prices.[31] To manage future changes in prices, allowances may be banked freely from year to year and from phase to phase; however, borrowing from future years and phases is not permitted.[32]

Within the RGGI framework, electricity companies could use up to 3.3 percent to fulfill their caps.[33] These will start functioning when the allowance reaches 7.00 USD. In this way, "[e]lectricity generators will be able to plan for and invest in lower-carbon alternatives and avoid dramatic electricity price impacts."[34] Up to 2012, the majority of bidders have indeed been electricity generators, accounting for 86 percent of the CO_2 allowances auctioned in Auctions 1–20.[35]

The regional cap is regulated by the CO_2-BTP, and it was established at 4 percent above the average annual emissions of 2000–2004.[36] For 2009, the limit was set at 188 millions of tons annually.[37] In other words, it is expected that electricity plants will emit in this time period a total cap of 188 million tons of CO_2. For 2015, this cap will be reduced by 2.5 percent each year until it reaches 10 percent in 2019. This regional CO_2 cap is divided among participating LGs, by previous negotiation based on their historic emissions trend.[38] For example, of the 188 million tons, the State of New York received 64.3 million tons of CO_2 as its limit. This means that New York would get the same amount of allowances as would be sold to its electricity plants.[39] In the updated Model Rule, it was established that the regional emissions cap starting in 2014 would be 91 million tons.

30 Farnsworth, 2010.

31 The price of the allowance for 2009 was 2 USD per ton of CO_2. See Regional Greenhouse Gas Initiative, 2010a, 2010c; see also New York State, 2010.

32 Consilience Energy Advisory Group Limited, 2009:203–5.

33 New York State, 2010.

34 New York State, 2010.

35 Regional Greenhouse Gas Initiative, 2013a.

36 Farnsworth, 2010.

37 Federal Energy Regulatory Commission, 2010.

38 According to David Farnsworth, attorney of the State of Vermont for RGGI, one of the most difficult topics during the RGGI negotiations was to determine the caps of every state. For states wanting to adhere to this initiative, emissions of CO_2 from 2000 to 2002 will need to be averaged to establish their potential cap.

39 New York State, 2010. For RGGI, one allowance equals one ton of carbon.

Performance and instruments: Prices, CO_2 reduction and consolidation

In 2008 and 2009 RGGI's carbon market offered the first auctions. During the process, some of the electricity plants participating in this carbon market achieved CO_2 reduction. They did not use all the allowances, so they were permitted to participate in secondary carbon markets. However, some of them needed more allowances, which could also be obtained in other RGGI auctions, offset programs, or in secondary markets. It is important to mention that secondary markets can come from other transregional initiatives, some recently created national markets or even the KP's market through its Clean Development Mechanism—even though the US is not part of it.[40]

RGGI can use the Chicago Futures Climate Exchange, the Montreal Climate Exchange, the European Climate Exchange and so on. In other words, RGGI allows versatility to its members to link with different levels of government, regional, national and global. To guarantee that all members can participate within RGGI or other markets, one of the main tools is accurate measuring instruments for emissions. RGGI has a database of CO_2 emissions of all of its member states. This database offers society the chance to know the reports and activities of the regional market, which grant legitimacy and transparency.

This database is called the "CO_2 Allowance Tracking System," best known as "RGGI-COATS." In this system, the way of measuring and reporting CO_2 has been harmonized. RGGI regulations for measuring CO_2 are the same as those used by the Division of Markets of Clean Air of the Environmental Protection Agency.[41] The reports that each state presents to the EPA are transferred directly to the RGGI-COATS, and can be consulted in both databases.[42] Although the EPA established guidelines for reports on emissions only at the federal level,[43] it recognizes that "the Regional Greenhouse Gas Initiative, which is the first mandatory cap and trade program in the U.S. for CO_2, is exclusively a State program."[44]

Besides assuring data transfer with EPA, the CO_2-BTP informs about allowances trade, establishes the basis for auctions, monitors the market for carbon, and gives advice to member states. This system informs each state of the CO_2 emissions of every electricity plant within each member state. Likewise, it monitors the offset system. The offsets projects are registered and verified by RGGI-COATS in each of the member states. Examples of offset projects are the recovery of biogas, methane or forestation. In this last category (forestry), a new modality was included in the 2013 updated Model Rule, the "Sequestration of carbon due to reforestation, improved forest management or avoided conversion." This modality was designed

40 Ellis and Tirpak, 2006:9.

41 Guidelines for the emissions report of each state can be consulted in Rule 40 CFR Para 75 in Environmental Protection Agency—US, 2009.

42 Regional Greenhouse Gas Initiative, 2010b and 2010c.

43 Federal programs that it regulates are the Acid Rain Program and the Clean Air Interstate Rule.

44 Environmental Protection Agency—US, 2009:3.

under the California Air Resources Board (CARB) US Forests Offset Protocol and it includes projects that will improve forest management, avoid conversion, and reforest areas under an afforestation condition.[45]

This performance ensured that RGGI become the first working carbon market in North America. The carbon market works through the reciprocity of the members. For example, regulated power plants can use a CO_2 allowance issued by any of the ten participating states to demonstrate compliance with the state program governing their facility. Thus a regional approach underscores each member's activities, and, when taken together, the ten individual state programs function as a single regional compliance market for carbon emissions.[46]

In spite of these efforts, results have not been as expected. As shown in Table 6.2, all members registered unfavorable results over time. This situation has led some members to question their participation in the market. This is the case of New Jersey, which, in November 2011, notified RGGI of its withdrawal from the market. The main reason for this drop in profit was the low prices. For the first phase (2009–2011), this carbon market offered 30.8 million CO_2 allowances at an average price of 3.23 USD per allowance.[47] However, the legitimacy of RGGI was questioned in the second phase where the price dropped to 1.86 USD per ton.[48] Thus

> the 188 million-ton cap set by RGGI is too high to have much effect. The cause appears to be an epic miscalculation in RGGI's projection of carbon emissions. In 2005, utilities released 184.4 million tons of CO-2. However, emissions plummeted by one-third—60million tons—between then and 2009, when the cap began.[49]

If we compare CO_2 emissions for 2009 with those of 2012, it is evident that, although the profits have not been there, there have been other types of incentives to continue. One of these has been that the price has increased. For 2011 it was 1.89 USD, for 2012 it was 1.93 USD per ton, and for the first half of 2013, the price reached 3.21 USD. Another incentive is the effective CO_2 decrease in the majority of members. As shown in Table 6.2, only four member states (one being New Jersey) increased their emissions; this increase did not exceed 24 percent in general. The rest of the members decreased their CO_2 levels. In addition, Table 6.2 portrays the number of plants that each participant has registered under RGGI. In spite of the drop in profit, the number of facilities participating has remained

45 Regional Greenhouse Gas Initiative, 2013b.

46 The Transportation & Climate Initiative of the North-east and Mid-Atlantic States, 2010.

47 Environmental Leader, 2010.

48 Potomac Economics, 2010.

49 Lagerkvist, 2010.

Table 6.2 Income of the RGGI states from participation in the "CO_2 Budget Trading Program" for 2009 and 2012

	2009			2012		
	Sold allowances (CO_2 tons reduced)*	No. of registered plants	Total revenue from auctions (USD)**	Sold allowances (CO_2 tons reduced)*	No. of registered plants****	Total revenue from auctions (USD)**
CT	7,322,363.83	21	17,686,118.10	6,819,154.547	18	11,401,368.85
DE	3,708,331.13	9	8,427,569.60	4,839,521.925	8	5,791,702.26
ME	18,661,075.54	31	9,618,943.79	13,217,639.587	27	5,507,366.94
MD	26,568,682.67	18	61,881,589.81	20,596,978.696	15	38,799,832.20
MA	3,643,492.68	6	50,918,303.38	2,940,072.416	6	28,530,599.42
NH	6,337,055.87	5	12,936,147.74	4,642,898.275	5	7,732,377.09
NJ	16,359,443.06	41	32,412,559.26	17,117,779.456	39	22,431,276.00
NY	37,700,660.56	83	125,430,330.07	35,417,900.566	71	65,707,560.24
RI	3,416,782.91	6	5,092,494.49	3,735,785.470	4	2,857,538.70
VT	1,965.30	2	2,347,481.42	2,319.100	2	1,317,244.30
Totals	123,719,853.55	222	283,288,011.75	109,330,050.04	255	190,076,866.00

Source: Author's elaboration with data from reports of each of the RGGI member states. For further details, refer to Regional Greenhouse Gas Initiative, 2013a.

* One allowance = One ton.

** Includes the four auctions taking place in each of those years.

*** New Jersey withdrew from the carbon market in November 2011 and did not participate in the 2012 auctions. Data in Table 6.2 for that year account for four auctions in 2011. This was included to compare the evolution of state revenue; however, it cannot be used to make comparisons with other states.

*** The number of facilities was counted by company. Some of the companies participate with different branches or subsidiaries and some of them do not participate in the market in certain years. However, if at least one of the branches participated, they were included in the table.

almost constant. In general, Table 6.3 shows the annual size of the carbon market resulting from trading emissions.

Table 6.3 Regional revenue resulting from auctions (2008–2012)

Year	Regional revenue in USD
2008 (2 auctions)	8,852,818.50
2009 (4 auctions)	19,686,118.10
2010 (4 auctions)	18,361,643.71
2011 (4 auctions)	8,865,753.69
2012 (4 auctions)	11,701,078.85

Source: Author's elaboration based on data from Regional Greenhouse Gas Initiative, 2013a. There have been 20 auctions in RGGI. Auction 1 was held on September 25, 2008 and Auction 20 took place on June 5, 2013.

The RGGI recognizes price uncertainty, and to address this problem it establishes mechanisms in the case of "trigger events" that might destabilize the allowance price for both phases. In the trigger-event scenario, more offsets (within the region or KP ones) could be used but only if the 12-month rolling average of the RGGI price is equal to or greater than 7 USD/ton for the first stage and 10 USD/ton in the second stage.[50]

In addition, the updated Model Rule proposed a cost containment reserve, which consists of a fixed quantity of allowances, in addition to the cap, that would be held in reserve, and are only to be made available for sale if allowance prices exceed predefined price levels. This reserve establishes those fixed prices at 4 USD in 2014, 6 USD in 2015, 8 USD in 2016, and 10 USD in 2017. Each year after 2017, the reserve price will increase by 2.5 percent.[51]

Integration of infrastructure: Production of electricity

In general, RGGI produces electricity with little carbon. That is why it has not been too difficult to incorporate these caps and new investments into their markets. As shown in Table 6.4, with the exception of the states of Delaware and Maryland, the majority of the electricity sources in the region are different from coal, highlighting the importance of nuclear energy as well as natural gas. This fact has impacted the cost of electricity in the region, causing it to be the most expensive in North America.

50 For stage one, price in 2005 dollars, adjusted annually to the US Consumer Price Index; for stage two it adds 2 percent.

51 Regional Greenhouse Gas Initiative, 2013b.

Table 6.4 Production of electricity by source (percentage) in RGGI member states (2009–2010)

	Nuclear		Hydro		Natural gas		Renewables*		Coal		Petroleum products		Price, kWh	
	2009	2010	2009	2010	2009	2010	2009	2010	2009	2010	2009	2010	2009	2010
CT	53.9	50.2	2	1.2	30.5	31.5	2.4	2.2	7.9	7.8	3.2	2.2	18.21	17.39
DE	0	0	0	0	30.7	50.9	3.1	2.5	63.9	45.6	2.4	1	12.17	11.97
ME	0	0	32.6	22.4	44.7	49.2	18.6	24.4	0.3	0.5	3.9	1.9	12.89	12.84
MD	33.3	32.1	4.5	3.8	4	6.6	0.8	1.3	55.4	54.2	2	1.1	13.11	12.7
MA	13.7	13.8	3.2	1.5	52.3	59.8	3.2	3	25	19.4	2.6	1.8	15.53	14.26
NH	44.2	49.2	7.9	6.7	26.7	24.2	5.6	5.6	14.5	13.9	1.1	0.3	15.2	14.84
NJ	55.7	49.9	0	-0.3	33	37.9	1.5	1.3	8.6	9.8	1.1	1	14.8	14.68
NY	32.5	30.6	21.1	18.2	31.7	35.7	2.9	3.5	9.6	9.9	2.2	0.6	15.66	16.41
RI	0	0	0	0	97.7	98	2	1.8	0	0	0.3	0	14.24	14.08
VT	72.5	72.2	22.2	20.3	0.1	0.1	5.2	7.3	0	0	3.8	0	12.75	13.24
National US (2012)	19		7		30		5		37		1			

Source: Author's elaboration with data from the Energy Information Administration—US, 2012a; National Mining Association, 2009b; Energy Information Administration—US, 2010a.

* Renewables include wind, solar, biomass, geothermal.

Up to 2009, the region used 43.1 million tons of coal to produce 55,400 million kWh (7.78 coal/kWh). The rest of the US states used 938.1 million tons of coal to produce 1,764,500 million kWh (1.88coal/kWh). These results supported the process of decarbonization that occurred in RGGI. According to the White Paper presented in November 2010 by the New York State Energy Research and Development Authority, in 2009, a decrease of 60.7 tons of CO_2 was registered.[52] For 2010, the decrease in the use of coal in the majority of the states is evident. As Rhead Enion notes, the decarbonization of the US economy rests on reducing oil, and especially coal. This process, he continues, is achieved by its substitution with sources such as gas and renewables.[53] As shown in Table 6.4, an increase in the use of natural gas has taken place, but not in the renewables sector.

There are several reasons for this. The most important incentive to decarbonize was the decrease in energy demand (specifically for electricity), when uncontrollable events such as the weather played an important role. However, the 7 percent decrease in demand of electricity from 2005 to 2009[54] occurred for two reasons. The first was the economic crisis of 2008, when a major part of sectors such as construction and housing reduced their activities considerably. The second one was the change from oil and coal to natural gas, a sector that experienced a dramatic fall in prices since the 2008 crisis. That has meant greater supply of electricity from gas-fueled power plants (relative to coal-fired plants), more investment in new gas-fired generating plants, less investment in coal-fired generating capacity, and retirement of existing coal-fired capacity, all of which have contributed to lower CO_2 emissions.[55]

Nonetheless, energy efficiency and environmental incentives can also be reasons for the CO_2 decrease, especially in electricity production. For example, up to 2009 the state of Massachusetts produced only 4,300 kWh, although it uses 9.9 million tons of coal, whereas the state of Maryland, using almost the same amount of coal (10 million tons), produced 24,000 kWh. In general, the RGGI performance in decarbonization of electricity production is the highest in the country. A clear explanation is the use of cleaner technology, especially hydropower. However, the use of nuclear power is very important for some states. Although the electricity produced is clean, some very pollutant chemicals are emitted during the process.

Another important factor related to electricity is regulation and operation of infrastructure. According to the Federal Energy Regulatory Commission, almost all RGGI states are regulated under the Northeast Power Coordination Council. This Council also includes the Canadian provinces of Ontario, Quebec and the Atlantics, but excludes states such as Maryland, Delaware and New Jersey—part of the Reliability First Corporation.[56]

---

52 New York State Energy Research and Development Authority, 2010.

53 Enion, 2013.

54 New York State Energy Research and Development Authority, 2010.

55 Stavins, 2012.

56 Energy Information Administration—US, 2008a.

RGGI states that do belong to the Northeast Power Coordination Council have infrastructure connections. An example of this is the electric corridor that is being developed between Maine and New Brunswick. The first part of this project includes the transmission of 1,200–1,500 MW to both states, based on wind energy and natural gas. Also, electric integration has the goal of including mechanisms for carbon capture and storage, as well as creating portfolios that account for renewables.[57] Another example of local electric strategy is Hydro-Québec. This company is in charge of providing most of the electricity in this region through its hydro-electric plants, with the state of New York its most important commercial partner.[58] It is important to mention that this company exports clean energy to the region, avoiding leakage. Hydro-Québec has different interconnections among RGGI states. This is the case of the interconnection of the continuous current of high voltage that has linked Hydro-Québec with New England since 1986.[59]

Historically, from 1999 to 2008, Canada has exported electricity to the US. In this way, integration of the electric sector among these countries has started in the area of the Northwest Atlantic. For the RGGI area, the Canadian provinces of Ontario, Quebec and New Brunswick export 37,000 GW.[60]

As presented in this section, RGGI operates based on a transregional governance approach, where rules and interconnecting infrastructures consolidate its mandate. RGGI has been the first carbon market to work without a national policy. However, it has generated criticism for its relatively modest reduction goals and over-allocation of emissions credits. Likewise, another critique has been the change in fuels from coal or oil to gas, which is still an important source of pollution. The most important challenge RGGI faces is the volatility of carbon prices. To face this situation, the initiative has created instruments that can manage uncertainty. Nonetheless, the most important challenge has been to keep the structure of RGGI solid.

However, RGGI has been working to integrate LGs' individual efforts under a transnational institutional architecture that offers mechanisms of linking to federal or global schemes. This TGER has accomplished some important goals, the most important one being to give positive signals to market creation and reduce CO_2 emissions. In the case of a federal policy, RGGI is prepared for this transition if the goals are comparable with those of the initiative.

The Western Climate Initiative (CA + QC): Starting Operations

The Western Climate Initiative (WCI) built on the Western Governors' Association and was in some areas already working on GHG reduction in the individual states or in regional schemes. This partnership started in 2003 with the West Coast Global

57 Government of New Brunswick, 2009.

58 Press Republican, 2009.

59 Dukert, 2009:170.

60 Canadian Electricity Association, 2009.

Warming Initiative (California, Oregon and Washington). In February 2007, the Southwest Climate Change Initiative (Arizona and New Mexico) joined the West Coast initiative to create WCI.[61] Some other jurisdictions (Utah, Montana, British Columbia, Manitoba, Ontario and Quebec) joined the project. Nonetheless, in November 2011, some of them officially withdrew, leaving WCI with California and Quebec as the only partners; British Columbia, Ontario and Manitoba argued that they were not ready for the Phase I compliance period (2012–2015).[62]

Before the withdrawal of the majority of the partners in 2012, WCI represented 13 percent of GHGs, 19 percent of the population and 20 percent of GDP in the US. The Canadian partners accounted for 50 percent of Canadian GHGs, 79 percent of the population and 76 percent of GDP,[63] having the largest coverage of any GHG regional initiative.[64] However, after 2011, California and Quebec, as the only members participating in the market, had very little impact on GHG reduction. Still, WCI's regional specific commitment is to reduce GHG emissions 15 percent below 2005 levels by 2020.[65]

Unlike RGGI, this initiative had a transregional scope from the beginning. It clearly went beyond national borders and transcended neighboring LGs (see Figure 6.3); it even had LG Mexican observers, such as Baja California, Sonora, Chihuahua, Coahuila, Nuevo León or Tamaulipas. The majority of these Mexican states already had a GHG inventory and registries developed under the EPA methodology, some of them funded by WCI and individual members.

As mentioned, WCI's regional commitment is to reduce GHG emissions 15 percent below 2005 levels by 2020.[66] This regional goal is lower than those established by some of its original members, for example California and British Columbia.[67] Some experts note that this would cause a "race to the bottom" regarding climate policies.[68] In other words, if an LG cannot comply with its own

61 The original five signatories were Governors Arnold Schwarzenegger (CA–R), Janet Napolitano (AZ–D), Bill Richardson (NM–D), Christine O. Gregoire (WA–D) and Theodore R. Kulongoski (OR–D).

62 New Mexico had even taken the case to court. The Local Department of Environment presented a lawsuit against Governor Susana Martinez because she was trying to put on hold the enacting of environmental legislation that would enable NM to participate in WCI. Although the court granted approval of the environmental enactment of rules for GHG management, the state will still be part of WCI but won't participate in the first auction. For details of this case, see Danish et al., 2011a.

63 British Columbia Ministry of Environment, 2010; Pew Center on Global Climate Change, 2009a.

64 World Resources Institute, 2010; Pew Center on Global Climate Change, 2009a. See also database of Department of Commerce—US, 2013, and for population data see Bureau of the Census—US, 2013.

65 Western Climate Initiative, 2007.

66 Western Climate Initiative Statement of Regional Goal, 2007.

67 Lee, 2011.

68 Brown and Sovacol, 2011.

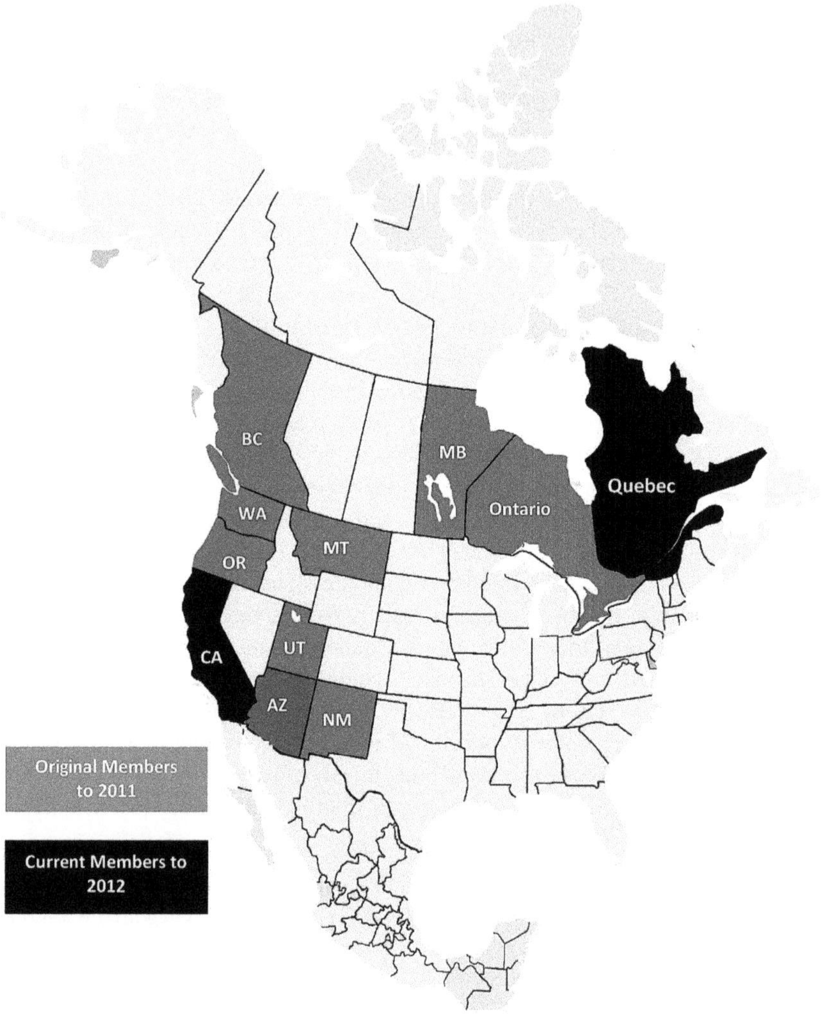

Figure 6.3 Geographic location of WCI members

Source: Adapted from Western Climate Initiative, 2012c.

domestic commitments, it can utilize the lower standards of the regional initiative. However, the logic of TGERs is meant to compensate regional inequalities regarding policy-building and implementation capacities, where if some states cannot comply with certain commitments, they can indeed rely on a wider, more flexible market. In this way, the regional goal would be achieved with market mechanisms and integration of infrastructure and policies, where harmonization and convergence perform within the region.

One of the most important features of WCI, which differentiates it from other markets, is a clear leadership to push forward instruments and policies. As established in Chapter 2, one of the causes of the failure of global climate-change governance had been the lack of true leadership without the commitment of the big polluter at the time, the US (and now of China). As in RGGI, leadership of certain LGs is fundamental for the carbon markets to operate. Thus California has been the leader of WCI since the beginning. In fact, the WCI program evolved from several of California's programs to reduce GHG emissions to 1990 levels by 2020, as well as from the California Air Resources Board (CARB).[69]

California's legal framework to address climate change was established in 2006 when the Global Warming Solutions Act (also called "Assembly Bill" or "AB 32") allowed CARB to decide on the focus of the mechanisms to be used (regulation or market based). CARB decided to use a mixed approach with regulations and a cap-and-trade scheme, where state-wide GHG emissions would be capped at 1990 levels by the year 2020. The cap also covered 85 percent of California's emissions and included the electricity, transportation fuels, natural gas and industrial emitting sectors. The trading of GHG allowances would involve electricity generators and large industrial facilities (with annual emissions greater than 25,000 metric tons of CO_2e). The cap-and-trade system includes offset programs within California and "outside" the state, where linking mechanisms to WCI cap-and-trade were designed.[70]

California has proven that leadership in transregional initiatives can be profitable, whether it is the 56 billion USD it has saved since passing groundbreaking energy efficiency standards in the mid-1970s or the 9 billion USD saved in clean energy venture capital that has been invested in the state over the last few years. If 23 of the leading policies being implemented in states went national, the US could reduce its carbon pollution to 27 percent below 1990 levels by 2020, far surpassing President Obama's short-term commitment of 17 percent below 2005 levels by 2020.[71]

When WCI was designed, California shared leadership with the Canadian province of British Columbia. In fact, in 2010, the province hosted the first carbon-neutral WCI meeting. To make this possible, the firm Pacific Carbon Trust (PCT) participated by granting offsets to the participants.[72] In general, WCI estimated to

69 Milne, 2008.

70 LaMotte et al., 2009.

71 Walker, 2010.

72 The Pacific Carbon Trust is a Crown corporation established in 2008 to manage British Columbia's GHG offsets, to help clients meet their carbon reduction goals, and to support the growth of the province's low-carbon economy. By 2011, the PCT delivered up to 1 million tons in carbon offsets to its public sector and corporate clients, making it one of North America's largest offset providers. For more information, see Pacific Carbon Trust, 2011.

save the region about 100 billion USD between 2012 and 2020.[73] It was thought that WCI was on the road to consolidation. Yet, as British Columbia withdrew the WCI, alongside the majority of its members, it lost the membership also.

Nonetheless, as mentioned in previous chapters, British Columbia had a mixed climate policy, in which it combined a carbon tax (since 2008) with the WCI cap-and-trade system. However, with the withdrawal from the WCI carbon market, the province retained the taxing mechanism. The British Columbia Clean Energy Act established a charge of 10 CAD per ton of GHGs in 2008, increasing by 5 CAD per year for four years, reaching 30 CAD in 2012. This tax is collected upstream (based on the carbon content of fuels). The law also established that the tax revenue would be refunded through tax cuts to business and individuals.[74] When the tax began operating in 2008, the province's GHG was expected to decrease approximately by 3 million tons annually as a result of the carbon tax.[75]

However, some experts noted that the carbon tax for 2012 was too low to influence people's behavior, being about 6 cents per liter of gasoline (when it was first introduced in the middle of 2008 it was 2.3 cents). In this way, this tax is lost in the swingeing price of gasoline in the province. In addition, up to 2011, the tax has had negative revenue, with the value of the tax cuts and credits exceeding the carbon tax receipts (with losses of 192 million CAD). Nonetheless, it is expected that the tax will close this gap in the long run.[76]

The remaining Canadian partner, Quebec, has consolidated its participation in WCI by establishing a close relation with California. On December 14, 2011 the Government of Quebec adopted the regulation respecting a cap-and-trade system for GHG emission allowances. One year later, this regulatory scheme was amended to harmonize Quebec's cap-and-trade system with California's, and those of possible partners such as Ontario and British Columbia. It also set out the rules under which Quebec would issue offset credits. The first year of operation of the regulation scheme started in 2012, but the whole of that year was spent in preparing the participants for the market. The regulated entities were required to register in the tracking system CITSS (Compliance Instrument Tracking System Service).[77]

The Quebec cap-and-trade system is based on recommendations of WCI original members since 2008, especially with the advisory of CARB. The Ministère du Développement durable, de l'Environnement, de la Faune et des Parcs is the institution that operates the system, being responsible for the creation and distribution of allowances, transactions, auction results and reserve sales. The cap-and-trade in the province has three compliance periods (2013–2014, 2015–2017 and 2018–2020). In these periods, the regulated entities

73 Walker, 2010.

74 Aldy and Stavins, 2011.

75 Ibid.

76 Schultz, 2012.

77 Government of Quebec, 2013a, 2013b. See also Mansell et al., 2013.

can comply by using allowance trading, early-action credits and offset credits (up to 8 percent) issued by the government of the province or by other governments with linking agreements.

By contrast with RGGI, WCI has allowed California and Quebec to decide on the sanctions for non-compliance. It is interesting to note that Quebec has established that, if a participant in the cap-and-trade system fails to comply in the last month of each of the three periods, Quebec's government can impose either a financial penalty or an administrative sanction equal to three-to-one emission allowances, in addition to the requirement to complete other instruments. The government may limit the transactions the entities make until they deliver results. Thus it is important that Quebec is addressing competitiveness issues by giving certain amounts of free allowances to a strategic sector, such as aluminum, chemicals, paper and petroleum refining, among others. However, the link with WCI-California gridlocks some of these polluting sectors. For example, for the second compliance period, no distributors of fuels are eligible for free allowances.[78]

The entities participating in Quebec's first period can be persons or municipalities operating a facility whose annual GHG emissions (excluding CO_2 related to biomass combustion) are greater than 25,000 tons of GHGs. Quebec also incorporates entities distributing within the province electricity that derives from imports. The second compliance period will include fossil fuels greater than 25,000 tons of GHGs.

In spite of the first-stage missteps, WCI started operations in 2012 with the integration of the California and Quebec cap-and-trade systems, becoming the second transnational carbon market in North America. Needless to say, the California system is more developed and has experienced three auctions already. But, although the scope of WCI has been reduced to a bilateral market, the institutional architecture is supposed to encourage some LGs to return to this mechanism if no national law is passed in the short term and the economy starts to recover.

The institutions created by WCI integrate policies and instruments under an administrative entity that sanctions the members, similar to RGGI. The following sub-sections address how WCI integrates policies, how the institutional architecture performs, almost on behalf of the California institutional setting, and the results that the initiative has obtained. In addition, the last section addresses the interconnectedness of the energy infrastructure, especially in the electricity sector.

The carbon market in practice: Policies, strategies and instruments

The first step towards implementing a cap-and-trade mechanism was to create an institution with legal capacity. In 2012, the initiative took the form of a corporation, very similar to the RGGI scheme.[79] Oddly enough, this new corporation is based in

78 Government of Quebec, 2013a.

79 The official name of the market is Western Climate Initiative, Inc. See Western Climate Initiative, 2012b.

Delaware, a situation that has raised serious arguments by civil society and local legislators, especially from California.[80]

Despite this situation, the transition into a corporation gave WCI the legal basis to manage auctions and give other services.[81] The WCI cap-and-trade program will cover emissions of seven GHGs (carbon dioxide, methane, nitrous oxide, hydrofluorocarbons, perfluorocarbons, sulfur hexafluoride and nitrogen trifluoride) from electricity generation, including electricity imported into the WCI region, industrial fuel combustion, industrial processes, transportation fuel use, and residential and commercial fuel use. As in RGGI, the process was divided into phases. The first phase of the cap-and-trade program was to begin on January 1, 2012. However, it was delayed until August 2012, covering emissions from electricity, electricity imports, industrial combustion at large sources, and industrial process emissions for which adequate measurement methods exist. In this first phase, only California's entities participated in the 2012 auction. The second phase will begin in 2015, when the program includes transportation, residential, commercial and industrial fuels not covered in Phase I. By including flexible mechanisms, such as banking, offsets and three-year compliance periods, the WCI cap-and-trade program can provide a strong and balanced approach to address GHG reduction in a cost-effective way.[82]

In Phase I, the cap was implemented by issuing a limited number of emission allowances to entities and facilities covered by the program (called "Jurisdictions' Allowances Budget"). Considering the two phases of the program, each partner will need to calculate their allowance budget, and then choose whether the allocation will be free or sold.[83] Thus the WCI cap-and-trade is composed of the aggregate of both individual jurisdictions' programs implemented through state and provincial regulations. This characteristic speaks of a balance between regulatory and market approaches. Nonetheless,

> the choice of allocation method [free or auctioned] [has] profound distributional impacts. There are potential gains for low-to-middle-income households if permits are auctioned and revenues spent in a progressive manner … However, existing high emitters should be expected to lobby hard for free permits based on current levels of pollution, and if successful they would reap an enormous financial windfall.[84]

80 Grimes, 2012.

81 Western Climate Initiative, 2010a:25.

82 Western Climate Initiative, 2010h.

83 Ibid.

84 See Lee, 2011:24. However, as some partner governments have withdrawn from the first phase of the market, California will need to give away most of its allowances to polluters in the early years of the plan to be included in participation. For further details, see Gardner and Henderson, 2010.

To deal with the interests of each LG's market participants, WCI established that whether allowances are free or sold would be addressed by each member. In other words, California and Quebec establish their own allowance budget. In general, free distribution of allowances is being granted to energy-intensive or trade-exposed industries based on benchmarking, and to the electricity sector (by a system of calculating imports into the WCI partners).[85]

The total number of allowances (the sum of each member) serves as the cap on emissions for the region. In other words, although each LG has the right to negotiate its goals and number of allowances, the general cap will be established for the region. The way WCI found to make emissions low and more expensive over time is that, "[a]t least once each three years, covered entities [e.g. companies] are required to turn into the state or province one emission allowance for each metric ton of CO_2e emissions they emit and report."[86] In the WCI scheme, entities that emit less than the cap can sell or hold their allowances. Selling excess allowances permits members to get back some of their emissions reduction costs. If they intend to hold allowances to use them later, this will only increase future compliance costs.[87]

WCI integration's logic is to sum individual members' efforts in a middle-and-around scope, where individual policies aggregate to form transregional linked policies. The most common of the individual (state/province) mechanisms developed by the members of WCI are GHG inventories, climate action plans, public benefit funds, inventories and registries, renewable portfolio standards and green pricing programs. Table 6.5 lists the most relevant mechanisms by members of WCI.

For WCI, the design of climate action plans is a necessary condition. These plans link individual members to the regional strategy. Members of WCI compensate each other through different instruments across many sectors in order to fulfill climate goals.[88]

Apart from the cap-and-trade itself, the most relevant elements that speak of a transregional approach are the CITSS and the reporting systems. WCI ensures the necessary institutional architecture for the regional carbon market to operate by creating tracking systems (CITSS) to guarantee the compliance of the cap-and-trade program. This system is used to (1) register entities participating in the California cap-and-trade program; (2) track the ownership of compliance instruments; (3) enable and record compliance instrument transfers; (4) facilitate

85 A benchmark is an "indicator of efficiency to compare the facilities or operations to an industry standard or best practice metric ... Use of common benchmarking would facilitate harmonization among similar facilities within the same industry addressing potential competitiveness concerns." See Western Climate Initiative. 2010a:14.

86 Western Climate Initiative, 2010a:6.

87 Ibid.

88 Pew Center on Global Climate Change, 2009a:3–7.

Table 6.5 Types of individual mechanisms of members of WCI

	Climate action		**Energy sector**			**Transportation sector**	
	Climate action plans (emission targets)	**GHG inventories and registries**	**Renewable portfolio standards**	**Public benefit funds**	**Green pricing and net metering (electricity)**	**Financial incentive for biofuels**	**Vehicle GHG and renewable standards**
				Members up to August 2012 (Phase I)			
CA	33% below 1990 levels by 2020	X	33% by 2010 (mandatory)	For renewables and efficiency	X	X	X
QB	20% below 1990 levels by 2020	X	100% (3% of wind power— 98% already comes from hydropower) (mandatory	For efficiency and renewables	X	X	X
				Original members			
AZ	2000 levels by 2020	X	15% by 2025 (mandatory)	For renewables and efficiency	X	None	X
BC[**]	10% below 1990 levels by 2020	X	13.4% by 2020 (mandatory)	For renewables and efficiency	X	X	X
MB[***]	6% below 1990 levels by 2012 (mandatory)	X	None	None	X	X	X
MT	1990 levels by 2020	X	15% by 2015 (mandatory)	For efficiency	X	X	None
NM	10% below 2000 levels by 2020	X	20% by 2020 (mandatory)	For efficiency	X	X	X

Table 6.5 cont.

	Climate action		Energy sector			Transportation sector	
	Climate action plans (emission targets)	**GHG inventories and registries**	**Renewable portfolio standards**	**Public benefit funds**	**Green pricing and net metering (electricity)**	**Financial incentive for biofuels**	**Vehicle GHG and renewable standards**
			Original members (cont.)				
ON****	15% below 1990 levels by 2020	X	67% by 2020 (mandatory	For renewables and efficiency	X	X	X
OR	10% below 1990 levels by 202	X	25% by 2025 (mandatory)	For renewables and efficiency	X	X	X
UT	2005 levels by 2020	X	25% by 2025 (mandatory	None	X	None	X
WA	1990 levels by 2020	X	15% by 2020 (mandatory	None	X	X	X

Source: Author's elaboration with data from Environmental Protection Agency—US, 2013; US states from Knigge and Bausch, 2006; Pew Center on Global Climate Change, 2009a.

* For data on climate change for Quebec province, see Government of Quebec, 2008. For data on energy and transportation sectors, see Government of Quebec, 2007. ** British Columbia is the only member to have a carbon tax (since 2008). For further details, see Aldy and Stavins, 2011; Duff, 2008. Data for British Columbia's strategies on climate action were taken from Government of British Columbia, 2008. For data on renewable portfolio standards, see Ethree.com, 2007a and 2007b. For data on net metering, biofuels and vehicle renewable standards, see Perkins Coie and Fasken Martineau, 2008. For vehicle standards data, see Government of British Columbia, 2002. *** See Government of Manitoba, 2008b. Most of Manitoba's power production is already from renewable sources. It is unlikely that the province wants or needs a renewable standards portfolio. Manitoba currently generates 5000 MW of electricity, over 95 percent of which is hydro from the Nelson, Winnipeg and Saskatchewan rivers and which reflects only half of Manitoba's hydro potential. For further details on this topic, see The Canadian Bioenergy Association, 2005; Government of Manitoba, 2008a. **** See Government of Ontario, 2008. For information about renewable portfolio standards of Ontario, see The Energy Roundtable, 2010. For data on the transportation sector, see Government of Ontario, 2007.

emissions compliance; and (5) support market oversight.[89] For example, WCI has standardized databases, individual member records that are public and consistent with other carbon markets. CITSS also deals with verification and enforcement processes.[90]

In Quebec, the Ministère du Développement durable is the body that accepts and verifies the identity of the entity, as well as grants permission to participate in the cap-and-trade system. Like RGGI, WCI has its own reporting system for GHG emissions, based on the methodology of The Climate Registry. This is so important for WCI that, if a new member wants to join the association, it has already to be a participant in The Climate Registry. To avoid overlap, reporting systems are harmonized and facilities can send almost the same report[91] to The Climate Registry, the federal Environmental Protection Agency of the US, and Environment Canada for the Canadian provinces.[92] Thus reporting is voluntary and applies to the seven GHGs specified in the KP, not only to CO_2, as RGGI does. These reports are the main instruments for participating in carbon markets. Also, WCI reporting allows this transregional initiative to participate in the Environmental Atlas of North America (CEC).

As the design program of WCI establishes, those facilities and companies covered by the cap may trade their allowance budget at regional auctions (where there is a purchasing limit). It is important to note that each auction has a price floor, which helps to correct over-allocation (where prices go below the environmental standards). A minimum price was set annually, starting in 2012 at 10 USD for California and 10 CAD for Quebec, both increasing by 5 percent annually (plus inflation) until 2020. If joint auctions take place between two WCI members, the highest price applies, according to exchange rates.[93]

WCI tries to prevent emissions leakage in both phases. This occurs when production activity shifts from WCI members to non-partner ones, so that emission reductions in WCI are transferred to other jurisdictions. WCI offers three compliance mechanisms that address the problem of leakage: early reduction allowances (for the period up to 2012), offset certificates, and approved "outside" mechanisms (such as participation in other carbon markets or secondary markets). However, the last two may not exceed 49 percent of the aggregate required emission reductions within WCI.[94] This flexibility is aimed to avoid overlap with

89 For details on how to participate in this system, see California Air Resources Board, 2013b.

90 Western Climate Initiative, 2010a:24.

91 Some of the reporting systems are not for carbon markets, but for other purposes. However, WCI asks for more details in order to comply with its own cap-and-trade mechanism. Canadian federal and provincial governments also work a joint reporting system.

92 See Western Climate Initiative, 2007 and 2010d.

93 Government of Quebec, 2013a.

94 The 49 percent emissions reduction "outside" WCI translates into a very small percentage when calculating the total emissions of the region. For example, if an

other policies and other carbon markets. It is also aimed to reduce compliance costs and allowance prices, especially when thinking of the interconnectedness of the energy sector in North America. In other words, LGs can buy and sell allowances on secondary markets (auctioneers can purchase a limited number of offsets for sources not covered by the cap), bank them for future use (store them to complement future compliance periods), or receive credits for early actions (if the entity proves to have reduced GHG and intensity emissions).[95]

WCI members agreed that a minimum percentage of the value of each member's allowance budget would be dedicated to energy efficiency and renewable energy incentives and achievement; research, development, demonstrations and deployment for carbon capture and sequestration projects; renewable energy generation, transmission and storage; energy efficiency; and promoting emission reductions and sequestration in agriculture and forestry and other uncapped sources. The remaining percentage would be distributed by members to reduce consumer impacts, to transit to energy efficiency, to create green economies and green jobs, and to adapt to climate-change impacts.

WCI recognizes that costs for mitigation policies can increase if technology develops more slowly than planned; weather affects energy production; or uncertainty affects emissions estimates for the short-term compliance periods (2012 and 2015). For these situations, WCI included allowance reserves, an instrument used by CARB to use allowances for different compliance periods.

WCI has created flexible instruments to match those of other markets (future national ones or global ones). The WCI proposal includes "work[ing] closely with federal governments to promote national and international action and ensur[ing] coordination among state, provincial, regional and national programs."[96]

In this way, offsets are considered to reduce GHG emissions with projects that are real, additional, permanent, verifiable and enforceable;[97] they need to reduce GHG emissions outside the cap. The offset program is a clear example of a transregional approach. First, it addresses the issue of linking with other regional initiatives, national ones, or even with the KP. Second, it establishes that "offset projects can be located in Canada, the U.S. or Mexico,"[98] meaning that the concept of region includes the *de facto* NAFTA members and observers. This emphasizes the argument that economic integration in North America could also be understood as transregional integration, especially for energy and climate-change topics. Linking programs with other initiatives or governments multiplies the opportunities to achieve cost-effective lower emissions, to expand the geographic

---

emitter of 100,000 $MTCO_2e$ uses them, only 5,000 offset certificates or approved outside mechanisms are permitted.

95 Government of Quebec, 2013a; Western Climate Initiative, 2010h.

96 Western Climate Initiative, 2010f.

97 For definitions of all of these conditions, see Western Climate Initiative, 2009a and 2010b.

98 Western Climate Initiative, 2010a:17.

scope to maintain competitiveness, to enlarge the size of the market, and to share operations costs.

Performance and instruments: Prices, GHG reduction and consolidation
The Phase I compliance of the WCI period started with the first auction in November 2012 among bidders from California.[99] The number of allowances has varied from auction to auction, dropping from the first to the third by almost 40 percent, as shown in Table 6.6.

Table 6.6 WCI auction results

	Allowances sold*	Bidders	Median price of allowance (USD)	Total revenue (USD)
Auction 1 (November 1, 2012)	23,126,110	73	11.81	233,342,450.00
Auction 2 (February 2, 2013)	12,924,822	91	12.91	176,036,076.00
Auction 3 (June 5, 2013)	14,522,048	81	14.25	203,308,672.00
Totals	50,572,980	245	12.99 (average)	612,687,198.00

Source: Author's elaboration with data from California Air Resources Board, 2013a.
* One allowance = 1 ton GHG.

In the same way as RGGI, the number of bidders has maintained and even increased.[100] This reflects the goal for a carbon market to be successful. In time, it was supposed that the number of allowances would drop and the number of bidders increase, to make allowances more expensive. This would encourage bidders to invest in technological changes or in new mechanisms to manage GHGs. The data presented in the three auctions show that this logic performed clearly from the first to the second auctions. As some analysts say, these results show a more robust level of market involvement than that of the first auction in 2012. In fact, the results would have been better if the credibility of the November auction had not been seriously undermined when CARB disclosed that one bidder had accidentally purchased 16 million CAD worth of allowances it never intended to buy. In general, the market

99 Up to the time of writing, it is unclear when Quebec will participate in its first auction. For details, see Mansell et al., 2013.

100 Quebec has 80 regulated industries; California regulates more than 300 industries.

seems strong.[101] However, this success had a slight drawback regarding the third auction, but it is expected that this market will consolidate in the short term and find robustness with the inclusion of the Quebec bidders.

However, contrary to RGGI, the prices at WCI had been rising consistently. As Enion notes, this price certainty gives stability to the market and motivates bidders to participate, anticipating costs and benefits.[102] Legislation passed one month before the first auction in 2012 acknowledged the need to invest the profit of the market in the social sector by establishing that 25 percent would go directly to public benefit funds for disadvantage communities.

Another concern is the actual GHG reduction. Results here are yet to be seen, but the choice of energy sources in the state has developed towards the same goals of the carbon market. California has been improving its GHG reductions and has been pioneering solar power development, as well as improving innovation engineering in Silicon Valley and funding research in world-class universities and national laboratories, while the state's politicians show a keen awareness of the climate-change problem, as demonstrated by the work of the Democrat party.[103]

Integration of policies based on clean energy sources: Production of electricity

Originally, WCI was based on the physical integration of energy sources in the region. As explained in previous chapters, this integration related to production and consumption of fossil fuels and renewables. Geographical factors were important for this integration, as in the case of RGGI. Usually, integrated infrastructure is located in neighboring states or provinces, with similar geographical and climatic diversity. In this way, WCI was thought to establish the importance of a broad geographical scope that would also reduce overall compliance costs and help mitigate leakage risks.[104]

However, with the withdrawal of the majority of the members, this geographical closeness lost its logic in WCI. California and Quebec are separated by thousands of kilometers. Nonetheless, the integration, not of infrastructure, but of policies, relies on the type of resources they use to fuel their economies. This section addresses an example of policy integration based on similar resources to produce electricity.

The LGs of California and Quebec belong to different electricity regions. California is inscribed in the Western Electricity Coordinating Council (WECC) and Quebec in the Northeast Power Coordinating Council (NPCC). Both councils regulate the production and transmission of electricity on a regional basis. Under WECC, California is divided into North and South regions.

However, thanks to the WCI regional effort, electricity emissions associated with imports can be incorporated into the offset schemes (either from both partners

101 Sloan et al., 2013; Poloncarz et al., 2013.

102 Enion, 2013.

103 Gerdes, 2013.

104 Western Climate Initiative 2010a:6.

Table 6.7 Production of electricity by source (percentage) in WCI member states (2009–2012)

	Nuclear		Hydro		Natural gas		Renewables*		Coal		8.2Petroleum products12.32		Price, kWh**	
	2009	2010-	2009	2010	2009	2010	2009	2010	2009	2010	2009	2010	2009	2010
Current members														
CA	15.5	15.8	13.6	16.4	55.4	52.7	12.5	12.5	0.1	1	0.8	0.5	13.58	13.01
QB	3	1.88	96	98.1	0.6	0	0.3	0	0	0	0	0	7.45	6.88
Original memebers														
AZ	27.4	27.9	5.7	5.9	31	26.6	0.2	0.3	35.5	39.1	0.1	0.1	9.57	9.69
BC	0	0	91	95.3	7	5.9	2	0	0	0	0	0	6.63	7.79
MB	0	0	98	98	0	0	0.1	0	0.1	0	0	0	7.4	7.08
MT	0	0	35.5	35.6	0.3	0.2	3.4	3.4	58.4	62.4	1.8	1.4	7.44	7.88
NM	0	0	0.7	0.6	21.8	23.5	4	5.1	73.4	70.7	0.1	0.1	8.2	8.4
ON	54	65	23	32.2	7	2.02	0	0	16	10	0	0	12.32	11.40
OR	0	0	58.3	55.4	28.5	28.4	7.5	8.6	5.6	7.5	0	0	7.63	7.56
UT	0	0	1.9	1.6	14.8	15.3	1.1	1.8	81.6	80.6	0.1	0.1	6.78	6.63
WA	6.4	8.9	69.8	66	11.5	10	4.8	6.4	7.2	8.2	0.1	0	6.63	6.66

Source: Author's elaboration with data from Energy Information Administration—US 2013a, Energy Information Administration—US, 2012a; Canadian Centre for Energy Information, 2011; Energy Information Administration—US, 2010c; National Mining Association, 2009b. Data for Canadian provinces from Canadian Electricity Association, 2012; Hydro-Québec, 2011; Environment Canada, 2008a; British Columbia Hydro Regulatory Group, 2009. For Canadian provinces the exchange rates from January 2010 and June 2013 were used to calculate the price of electricity in USD for 2009 and 2012.

* Renewables include wind, solar, biomass, geothermal.

** Residential average prices in USD.

or from others), in spite of not being physically connected.[105] This means that regulation goes indirectly to the first entity that delivers electricity. Imports allowances can be stored in reserve in order to cover other compliance periods.

The distribution of electricity in the western connection accounts for "cap-and-trade energy" to be distributed through the California transmission lines (either imported or exported). The same happens with Quebec and its network. For example, the western interconnection links the grid to some of the Northern Mexican cities in the states of Baja California and Chihuahua, with electricity amounting to 1,000 MW.[106] California is the largest net importer of electricity, consuming power produced in the Northwest and Southwest. These two regions provide about 25 percent of California's electricity supply, with the requirement that a percentage of this meets the GHG reduction targets of the state.[107]

In the original WCI, the majority of members performed their activities with less coal generation in comparison with the rest of the country. None of the top ten carbon-intensive states belonged to WCI. It is understandable that withdrawal from WCI started with the coal-intensive states, such as Arizona, Utah or New Mexico. Table 6.7 shows the sources used to produce electricity in the original and current members of WCI.

When Quebec integrates fully into the auctioning process, it will be able to compensate for California energy production. As shown in Table 6.7, Quebec gets almost all of its energy from hydroelectric sources; the remainder comes from oil and natural gas. California has a balance between hydro, nuclear and renewables, but the true challenge is to offset natural gas production of electricity in other markets.

WCI's efforts towards decarbonizing the economy of its original members have come to lose logic with current members. However, California has taken advantage of some of the federal legislation enacted since 2007 regarding coal. In that year, the US Supreme Court determined that CO_2 as a pollutant should be regulated by the Clean Air Act and the Environmental Protection Agency. This piece of legislation created the Clean Coal Power Initiative Program, which would focus on capture and sequestration of GHGs. By 2009, it had allocated a total of 408 million USD to two projects—the Basin Electric Power Cooperative's Antelope Valley Station in North Dakota and the Hydrogen Energy Project in California—to demonstrate the capture capability of 3 million tons of CO_2 per year.[108]

As this section has suggested, the relation of WCI with the institutional architecture of California emissions agencies and process is profound. In some areas, it can be perceived that California's schemes have simply been transferred to the other LGs. One example is that the cap-and-trade system (auctioning, offsetting, monitoring etc.) is operated from CARB offices. The unclear relation

105 Western Climate Initiative, 2010a:15.

106 North American Energy Working Group, 2006.

107 Energy Information Administration—US, 2013b.

108 National Mining Association, 2009a.

between CARB and WCI has provoked comment. In the words of Diane Harley, "[i]nitially, the state [California] was to unite with other Western states to reduce the purported menace to the future of our planet. However, our partners determined that they would prefer not to tackle the issue during a recession; the cost of making their states less competitive in a tough business environment outweighed the benefit."[109]

Although this TGER is only just starting operations, it has been built on solid California legislation. The challenge for WCI is to demonstrate its viability, attract back the members that could not participate in the carbon market and persuade the observers to enter as full members. The only way to do this is to show results in terms of profit and low-cost technological change.

This chapter has presented the two operating carbon markets in North America, RGGI and WCI. Although different in scope, membership and instruments, both markets operate under the cap-and-trade system. RGGI has been successful in addressing its goals of managing CO_2 from power sources. WCI has successfully started the auction process in one of its partners. Both initiatives have addressed, in one way or another, the condition of having a transregional focus. RGGI has done so by establishing links with other types of markets, the European and the KP flexible mechanisms. WCI has linked two non-neighboring LGs that are part of different federations. It is evident that these initiatives are not perfect institutional architectures; they are always in the making and facing society's preferences and political change. They are dynamic. Needless to say, the market approach taken by them has been subject to criticism. The major challenges and opportunities will be summarized in the Conclusions to this book.

Now that RGGI and WCI have been introduced, Chapter 7 will present a comparison of the two transregional schemes regarding their goals and the flexibility they portray. Flexibility will be understood as the depth of the links the TGERs have designed: starting from convergence, following with harmonization and ending with linkage.

109 Diane Harley was interviewed by CalWatchdog in June 2012, when the market had not started operating. For further details, see Grimes, 2012.

Chapter 7
Linking RGGI and WCI to Other Climate Governance Schemes

Introduction

The two transregional initiatives described in Chapter 6 have similarities and differences among their members. This chapter shows how RGGI and WCI have developed linking mechanisms with each other and with other authorities (national, global) that favor compliance, as well as how similar the policies and their instruments need to be to deliver outcomes. In some cases, both TGERs exhibit deep convergence, at some other times, standardized processes, and at yet others, just superficial linkage. These three levels of linking are compared under specifically designed categories. This comparison is seen to be guided by radial concepts.[1] The design of radial concepts implies that the essence of the concept to compare is integrated by a variety of conditions. A radial concept integrates its "radiuses" (characteristics) and transforms them into the essence of a new concept. In other words, a radial concept is based on the connection of attributes in different degrees (or "diminished subtypes").[2] As Kurt Weyland puts it, "by relaxing the demands for definitional inclusion, a radial concept makes multidomain notions more broadly applicable and thus reduces the danger of false negatives."[3] For example, the concept of TGERs has been treated in this book as a radial concept with essential characteristics intrinsic to them: LG driven, managing GHGs with market approaches, and linking their economies to energy and environmental issues. Both TGERs have transregionalization processes going on with different forms of linking among their members and with others "outside" the region. In this way, regionalization provides the basis for establishing comparative criteria (secondary concepts). These concepts are convergence, harmonization and linkage. For each of them there is a brief explanation of: (1) approaches to environmental policy and institution-building; (2) policy goals; (3) policy instruments; (4) policy outcomes; and (5) infrastructure. Figure 7.1 helps to visualize this methodological scheme.

These variables represent different criteria under which an evaluation of RGGI and WCI can be made. However, this methodology is based on qualitative

1 Collier and Mahon, 1993.

2 Weyland, 2001.

3 Ibid.:3.

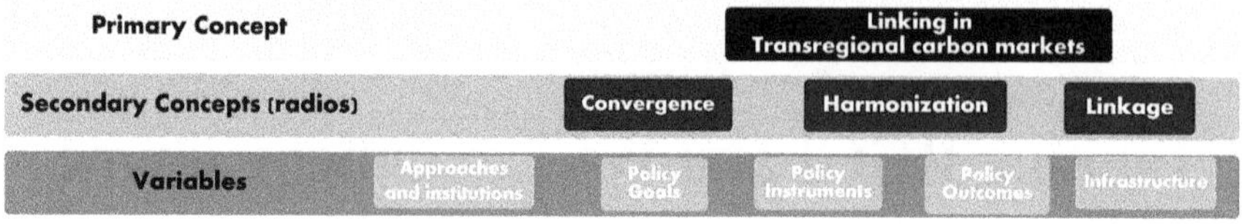

Figure 7.1 Comparing transregionalization

Source: Adapted from Van Nijnatten, 2008b; Harrison, 2007; Schwanen, 2006; Winfield, 2002; Moss et al., 2001.

interpretation of results presented in Chapter 6, and it is far from representing a quantitative exercise. The first section of this chapter includes the analysis of variables within the concept of convergence for both initiatives. The second deals with harmonization and the third with linkage. In the final section, the three criteria are compared under categories of high or low levels of linkage. High level means that more than half of the policies and instruments are similar in each of the carbon markets; low means that this condition is not met.

Convergence in Climate Transregional Policies

As defined in Chapter 1, "convergence" can be understood as the tendency of policies to resemble each other in terms of their structures, processes and performance. Convergence can be better analyzed by using some of the five variables in the scheme above (Figure 7.1). To deal with the approaches to environmental policy in TGERs it is necessary to address the main driver for managing climate change. The sharing of values is present when constructing an idea of GHG mitigation. This main idea in both TGERs is the recognition of a market-oriented approach through the creation of carbon markets and cap-and-trade systems. Put differently, "trading allows the convergence to the cap to occur in a cost-effective way."[4] This idea has also been recognized by other authorities, such as the flexible mechanism of the KP, and other carbon markets. The majority of the proposals for a national climate law in Canada and the US also used this approach. For example, this market approach acknowledges that convergence within TGERs is based on the coherence of economic activities, resources to power those activities, and mitigation goals. The approaches to environmental policy are also constructed by the principles that form the institutional architecture. One of the main principles recognized in carbon markets is that GHG emitters need to take responsibility for their pollution and are required to meet specific emission reduction targets. This vision is also shared by the KP and different mechanisms and by almost all national climate schemes.

4 Aguayo, 2010:248.

Another approach shared by TGERs is that, apart from being market solutions, transregional carbon markets are conceived as building blocks for national or global climate-change policies. This idea deals with the lack of efficiency in drafting policies at other levels of government.[5] The KP was based on national climate strategies and, except for the European Union, no country has built its national strategies on local initiatives.

Another important shared idea between the TGERs is leadership, whereby each region has its own "paymaster." As mentioned in Chapter 1, the provision of global public goods needs to have a leader who absorbs most of the costs, but gets most of the profit. Leadership is in the hands of the US LGs, as "only the American state participants are likely to have any prior experience with emissions trading under the 1990 Clean Air Act ... Neither Canada nor Mexico have this level of familiarity, at either federal or sub-federal levels."[6] California has promoted WCI (with British Columbia and Quebec); New York has led RGGI.

In the market–environmental approach common to both TGERs, "institution-building" is a cornerstone. Institutions of TGERs are designed to be flexible to allow change,[7] as the cap-and-trade mechanisms show. Institutional convergence within the initiatives is strong (especially in RGGI); however, it is not that strong in the face of other initiatives. At one point it was thought that WCI would converge directly with RGGI, transforming all structures into the same format. However, when WCI suffered withdrawal of the majority of its members, this idea vanished. Still, WCI has adopted the same legal form as RGGI and become a corporation with a phased approach. Yet convergence between these initiatives will not take place in the short run. In fact, WCI and RGGI have set two different models for establishing relations among their members. WCI has a more compact approach, where the jurisdictions are more autonomous and link with other climate policies only to create markets; RGGI has a more centralized approach, where jurisdictions share the design of market institutions. Prices of allowances are another example of the lack of convergence. In RGGI, prices have ranged from 1 USD to 4 USD, whereas in WCI (California auctions) the prices have exceeded 11 USD. Comparing the three first auctions of each market, RGGI accounted for 819,255,595 USD, and WCI for 612,687,198 USD. It is necessary to note that RGGI held these first auctions in 2008 and 2009, when the economic crisis was downsizing different sectors. These amounts of revenue were the result of the participation of 216 bidders in RGGI and 245 in WCI. Thus WCI has been more successful.

If convergence in institution-building has not been strongly present between TGERs, it is entirely absent when compared with national or global policies. The mechanisms of institutional architecture proposed by the KP are far from

5 Winfield, 2002.

6 Rabe, 2010:6.

7 Hall, 2003. See also Hall and Thelen, 2009. Kathryn Harrison also uses institutions and shared ideas to compare climate policy in Canada and the US. See Harrison, 2007.

similar to those of TGERs in various ways. The fact that the KP did not take into account economic activities, sectors and regional integration is the main obstacle for converging institutions. This same argument applies to the national policies proposed, which do not take into account regional differences.

This discussion leads to the analysis of "policy goals." As defined by Van Nijnatten, policy goals refer to "the stated or expected ends of the course of action to which a government is committed."[8] The expected goals are expressed in a specific time and place. In this regard, each of the TGERs has established its own goals for reducing GHGs. In fact, each of the LGs has done the same within the regions, where the sum of the parts performs as a grand regional goal. In this way, convergence among LGs' policies needs to be present. However, convergence among RGGI and WCI is weak: each set different regional goals to reduce specific GHGs in different time frames. For example, RGGI has committed to stabilizing emissions for 2014 and to reducing them by 10 percent for 2018 in 200 electric plants that use fossil fuels and generate more that 25 MW, starting on 1 January 2009. WCI aimed to reduce 15 percent of emissions below 2005 levels by 2020.

The scope of policy goals is another issue where there are differences. RGGI addresses CO_2 sources of only very few facilities, while WCI addresses the whole range of GHGs (as established in the KP) in a wider scope of sectors. Thus WCI uses the same scope as the KP and will be more prepared to link with national initiatives.

The sector scope is the one example where almost all authorities worldwide converge. In general, electricity is one of the most emitting sectors in the world. Thus transregional initiatives, other carbon markets (such as the European one), the KP and national policies agree to put it at the center of the discussion. Of the TGERs, RGGI has more interest in addressing electricity production to mitigate GHG emissions than WCI. If comparing energy portfolios, is easy to see that, in RGGI, between 15 and 30 percent of the energy devoted to electricity will be required to come from renewable sources. In WCI, California has set the goal of 33 percent and Quebec of 100 percent, taking into account that 98 percent of its electricity already comes from hydropower. Once again, the challenge for each WCI partner is totally different: for California it is to balance the energy mix and for Quebec it is to cut emissions by being more energy efficient and addressing other sectors (such as transportation).

To operate policy goals, "policy instruments" are needed. In RGGI and WCI, policy instruments are divided into three main areas: climate change, electricity and transportation. Although the pool of policy instruments tends to be very similar among all members of RGGI and WCI, they account for a low degree of convergence between the transregional initiatives. As presented in Chapter 6, each of the LGs adopts the most suitable instrument: vehicle standards, incentives to use biofuels, renewables, portfolio standards and so on. When comparing these instruments with other climate strategies, there is no common ground. For example, the KP mechanisms have nothing in common with the policy instruments

8 Van Nijnatten, 2008b:3.

chosen by LGs. Indeed, these mechanisms do not even resemble the cap-and-trade system of the TGERs.

The expected "policy outcomes" are more ambiguous and difficult to measure. No similar outcomes will emerge from the transregional approach compared with national and global schemes: not in institutional terms, not in efficiency use of energy, and certainly not in absolute GHG emissions reduction. But in general it is expected that any transregional initiative will accomplish three conditions: a sufficient degree of institutional robustness (solid institutions that can survive political change); institutional efficiency (high rates of compliance and future convergence); and positive environmental impact (GHG reduction and society's change in consumption and production choices).

RGGI is institutionally robust and efficient, but WCI, although efficient, has not been able to manage change and external factors, and has lost robustness. Do policy outcomes converge between the initiatives? The short answer is no. However, regarding environmental impact, both regions have converged

Table 7.1 Energy consumption per capita for members of RGGI and WCI (million Btu)

	RGGI	
	2010	2011
Connecticut	210	206
Delaware	284	299
Maine	306	310
Maryland	256	244
Massachusetts	213	211
New Hampshire	224	221
New Jersey*	278	276
New York	192	185
Rhode Island	187	175
Vermont	235	238
	WCI	
California	209	208
Quebec	144	156
US	315	312
Canada	205	204

Source: Author's elaboration with data from Energy Information Administration-US, 2013c; Statistics Canada, 2011.
Note: Btu = British thermal unit.
**Year* 2010 was the last year that New Jersey participated in RGGI

at least in the reduction of CO_2, accounting for 109,330,050 tons in RGGI up to 2012 (in 20 auctions) and for 50,572,980 in WCI (taking into account all GHGs) up to the same year in three auctions. The fact that the same bidders have been constantly participating in the auction process of both TGERs exemplifies a change in terms of production choices. Consumers within TGERs have also had a positive impact on consumption patterns by reducing their energy consumption per capita from 2010 to 2011 (Table 7.1). In both initiatives, the LGs have lower consumption per capita than the national averages of Canada and the US.

Finally, convergence appears to be a difficult concept to apply to the TGERs' infrastructure. To talk about convergence implies that all the energy production facilities, electricity lines or transportation routes would look much alike. As explained earlier, RGGI is interconnected physically with a network of transmission lines and a variety of pipelines. However, this infrastructure does not operate under the same technical standards or characteristics. In WCI there is no infrastructure converge at all. Thus convergence has more to do with the spirit of markets and sharing of general principles of differentiation of responsibilities within a regional scheme. Convergence has not reached these operative levels of carbon markets.

In general, isolation seems to be the common practice of these two markets, with little convergence. Faced with the intention to create a national carbon market, these two TGERs need to think about creating mechanisms to easily converge with each other to facilitate negotiation with a new federal proposal (either in Canada or, most likely, in the US).

Harmonization in Climate Transregional Policies

The fact that cap-and-trade is designed on the premise of cooperation in a context of differentiation of responsibilities and capabilities leaves no room for convergence, but opens the door to harmonization. Thus harmonization represents a less deep linkage than convergence. Harmonization can be defined as standardization. In the case of the cap-and-trade schemes in North America, harmonization allows emissions trading in several directions. As some scholars note,

> [u]nder a unilateral link, entities in system A can purchase allowances from system B for compliance, but not vice versa … inter-system trading will occur until prices of the two systems converge at an intermediate level. In a fully bilateral link, allowances can be freely traded between two systems and each system's allowances are equally valid for compliance in both systems. If more than two schemes participate, this becomes a multilateral link.[9]

9 Tuerk et al., 2009:343.

In any of these cases, cooperative governance is necessary to harmonize rules within TGERs, and with others. This means that certain norms should be applicable to all. As established earlier, there is convergence in the approaches of both TGERs, but is in the area of institutions and policy instruments that harmonization is at the core.

Institutions play a decisive role in harmonization. RGGI and WCI have developed institutions precisely for harmonization purposes. There are the RGGI-COATS and the CITSS for WCI to help their members to reach the regional policy standards. These two institutions give certainty to LGs in the long run and are registered in a common LG database called The Climate Registry (TCR). TCR emerged as a common reporting platform to create common trading and record-keeping databases to avoid double-counting. At the beginning, TCR was a voluntary program, but RGGI and WCI have used this platform consistently for the climate policies of each LG and for the regional schemes. As some authors state, TCR is has a "quasi-official status" for GHG reporting.[10]

Both TGERs have used harmonization as a strategy to prevent uncertainty due to a future national carbon market.[11] For example, WCI expresses the need to "develop administrative systems and infrastructure, continue advancing core policies and programs, work closely with federal governments to promote national and international action and ensure coordination among state, provincial, regional and national programs."[12] TGERs had harmonized their registries and inventories, not only within the regional schemes, but with the US federal EPAs, by following its methodologies.[13] Thus EPA is working with TCR to harmonize reporting and record-keeping practices among jurisdictions. In fact, RGGI has taken the first steps to harmonize with global governance institutions.

Harmonization, in the first place, operates in a multilateral scope (linking members within regions or allowing them to link globally). Second, it operates at a bilateral level by linking both TGERs under TCR, and third, in an indirect way, by linking TCR and EPA. However, harmonizing standards (and not converging) is not a random choice of these new markets; it was a precaution to avoid distributional issues, to attain of prioritized policy objectives, and to reduce control over domestic markets. The flexibility these markets show is designed to avoid overlap or double-counting, as well as to cut costs of implementation.

Another strategy to deal with uncertainty has been the creation of institutions that establish harmonized standards among regions, especially in offset programs and secondary markets. In May 2010, the three existing markets at the time (RGGI, WCI and the extinct MGGRA—Midwestern Greenhouse Gas Accord)

---

10 Freestone and Streck, 2009.

11 Rabe, 2010.

12 Western Climate Initiative, 2010f.

13 This happens in the case of Canadian and US LGs, where harmonization mechanisms have been created to homologate registries with the EPA. See Western Climate Initiative, 2010d and 2010c.

launched a project towards this end.[14] The offset system among TGERs was institutionalized by the creation of the Three-Regions Offsets Working Group.[15] In this way, harmonization was based on mitigation strategies, such as GHG reduction projects of sources not regulated in the cap-and-trade program.[16] The standards established by the three regions stated that compensation projects need to be real, complementary, permanent, verifiable and enforceable.[17] In other words, it was necessary that they made more GHG reductions than specified by the cap. Harmonization was multilateral: within each of the TGERs, standards were developed to cope with external offset projects from other regional initiatives.[18] However, there was a limit to offset projects outside the region. In the RGGI framework, "power plants might use offsets to meet 3.3% of their compliance obligation. Limit on use increases to 5 percent and 10 percent under specified conditions."[19] In WCI, the offset programs could not exceed the 49 percent of the reduction commitment of every member.[20] In that time, MGGRA created the Regulation Agency Office to decide on the offset projects to be accepted.[21]

The Three-Regions Offset Working Group facilitated the creation of systems to allow LGs to develop projects in secondary markets.[22] For example, the Market Committee of WCI established three types of harmonization mechanisms with secondary markets: exchanges, central limit order books (CLOB) and over-the-counter (OTC) transactions.[23] These three formats allowed flexibility and harmonized standards, but they are far from constituting convergence.

RGGI shows another example of how secondary markets can operate under TGERs. For RGGI, secondary markets can be other TGERs, where there can be common standards. RGGI can use the Chicago Futures Climate Exchange (national institution) as an exchange institution, as well as the Montreal Climate Exchange.

The Three-Regions project evolved into the North America 2050 project in 2012. The goals of this project are to: (1) coordinate efforts to design, promote and implement effective and cost-effective policies; (2) advocate the most appropriate roles for federal, state and provincial governments; (3) achieve meaningful emission reductions; and (4) demonstrate the economic and job creation benefits of policies. Even after the disintegration of MGGRA and the major withdrawals from

14 Rabe, 2010:24.

15 For the three regions, one allowance equals one CO_2e ton (CO_2 for RGGI). See Three-Regions Offsets Working Group, 2010.

16 See ibid.:6.

17 For details on the definitions of these characteristics, see Western Climate Initiative, 2009b and 2010b.

18 Western Climate Initiative, 2010a.

19 Regional Greenhouse Gas Initiative, 2011a.

20 This percentage translates into a very small amount if total emissions of the region are calculated. See Western Climate Initiative, 2010a:38.

21 Midwestern Accord, 2010a.

22 Western Climate Initiative, 2010a:17.

23 Western Climate Initiative, 2010e.

WCI, the members of the TGERs decided to create this institution as a platform for LGs in the US, Canada and Mexico, and as a transnational North American institution in case they wanted to participate in operating carbon markets. The NA2050 is divided into working groups, as shown in Table 7.2.

Table 7.2 NA2050 working groups

Working group	Goal
Benefits	Evaluate the benefits of policies and programs for moving to a low-carbon economy
Power sector	Understand, influence and participate in Clean Air Act requirements for electricity generation
Industry	Develop benchmarking approaches to encourage industrial energy efficiency, as well as preparing for Clean Air Act requirements for refineries
Sequestration	Explore regulatory, technical and policy issues surrounding carbon capture and sequestration
Sustainable biomass	Understand and support the use of sustainably harvested biomass fuels that have the potential to displace high-carbon fuels used by stationary and industrial sources
Offsets	Participate in the development and implementation of high-quality offsets that may be used in emissions trading programs or for other purposes
Linking	Work on issues related to administrative and policy linking of emissions trading programs as needed in the future

Source: Author's elaboration with information from North America 2050, 2012.

In general, these working groups cover the majority of areas in which carbon markets operate. In addition, this project will aid some economic sectors to conform to some federal decisions, such as ACES (described in Chapter 4).

In spite of the institutional harmonization efforts, specific policy goals and outcomes tell a different story. As mentioned earlier, although both markets share some approaches, the policy goals and certainly the outcomes represent substantial differences where no harmonization process has developed. Even though the policy goals are specific to each LG context, it is in policy instruments that a high level of harmonization is present. For example, California is entitled to set stricter standards than the nation as a whole and is exempt from weaker federal standards.[24] This process allows other states to do the same. In this way, New York has adopted

24 California is the only state whose vehicle emissions standards preempt federal rules. See Department of Environment of New York, 2011.

some recent California standards, for example to reduce GHG emissions from cars by 37 percent, and from light trucks by 24 percent by 2016.[25] This individual initiative has led towards cooperation among RGGI and WCI. Another example is the technologically neutral emissions performance standard being developed by RGGI and the State of Pennsylvania (not a member of any TGER). This standard reduces the carbon concentration in fuels used in vehicles and buildings, and provides incentives for energy providers to use low-carbon fuels.[26]

In the analysis that McKinsey & Company present every year regarding the main areas for GHG reduction and abatement costs worldwide, they show that, for the US, there are several strategies and instruments that need to be developed integrally. Harmonization of prices of energy (especially renewable) would be required to address the costs. In addition, harmonization of the instrument portfolio within each carbon market and among RGGI and WCI would also abate costs.[27]

The last of the criteria is infrastructure, where there is some evidence of harmonization, especially in the electricity sector, for example the expansion of electricity lines within the electricity regions already established and regulated by NERC. However, this harmonization causes overlap in the administration and provision of power to RGGI and WCI. However, the majority of infrastructure interconnectedness is found in the linkage area.

Linkage in Climate Transregional Policies

Linkage works in the same way as harmonization (unilateral, bilateral or multilateral).[28] For the case of North American TGERs, linkage is present with "outside" markets and both federal governments (Canada and the US). This linkage has been only unidirectional.

Linkage is a perfect strategy in the face of global governance failure. As established by several scholars, in case the world does not reach consensus about a global carbon market resulting from a new Kyoto scheme, national or regional markets are the only solution; thus they need to create linkage between their mechanisms to make them efficient.[29]

As established in the former subsections, the approaches to address climate change through carbon markets converge between TGERs and with certain areas of the KP. Unilaterally, RGGI allows its members to offset emissions in other climate-change schemes, such as the KP. Although unidirectional, this idea of

25 Ibid.

26 Ibid.

27 McKinsey & Company, 2007.

28 World Business Council for Sustainable Development, 2007.

29 An interesting proposal for linking carbon markets worldwide is presented by Flachsland et al., 2008. Some other scholars who propose linkage as a strategy for bottom–up construction are Victor, 2007; Pizer, 2007; Tangen and Hasselknippe, 2005.

compensating outside the cap with other instruments is present within the KP's flexible mechanisms.

Convergence or harmonization is not generally present in policy goals. Linkage is hardly present, either. However, when WCI had just begun operations, it intended to go further and establish multi-lateral linkages for allowance trading with the EU ETS and other trading schemes. As some analysts note, in late 2008 Governor Schwarzenegger hosted a summit with various LGs to explore these linking relationships.[30] This meeting was held within the Governors' Global Climate Summit, resulting in an agreement between US governors and 14 states and provinces from Brazil, Indonesia, Mexico and Nigeria to reduce forestry-related GHG emissions.[31] The group of LGs still participates in the UNFCCC, but no linking trading mechanisms have been put in place.

Linkage occurs in the area of policy instruments. For example, both initiatives recognize as valid some national or global instruments. RGGI is prepared to link to the ETS, to new national markets, or even to the Clean Development Mechanism of the KP.[32] This situation seems odd since the US federal government is not part of the KP. In other words, RGGI allows flexibility to its members to link with other levels of climate governance. Nonetheless, this link is far from joint institution-building or standardization. To have these links, is a precaution or accepted advantage for LGs, taking into account their potential inclusion into governance schemes, either national or global.

An example of linkage is the unilateral recognition of external credits by RGGI. This program allows participants to hold this type of credits outside the US in order to fulfill offset requirements whenever the price of carbon exceeds 10 USD. Another example of linkage is California's Scoping Plan under AB 32, which allows LGs to enter into offset agreements with foreign electricity providers to reach performance standards. WCI has similar schemes to recognize offset credits from other North American markets and the CDM.

Finally, linkage is the variable through which there has been more progress regarding regional integration of infrastructure. In the three TGERs, a diversity of fuels is used to produce energy. When the regions incorporate their sources into the general grid, they are in fact incorporating it into energy portfolios focused on managing GHGs. For example, "in the U.S., the emergence of Renewable Portfolio Standards (RPSs) is creating a greater need for access to Canada's renewable generation. An RPS requires electricity providers to obtain a minimum percentage of their power from renewable energy resources by a certain date."[33] As established in a Canadian report about infrastructure opportunities,

30 Andrews Kurth Attorneys, 2010.

31 Environment News Service, 2008.

32 Ellis and Tirpak, 2006:9.

33 National Energy Board—Canada, 2009a.

> additional transmission will give regions that are currently dependent on fossil fuel generation access to generation sources that emit less GHGs, such as nuclear, hydro and wind power. Wind power in particular benefits from transmission infrastructure. Areas with good wind resources may be located away from load centres, requiring transmission to bring the power to market. Locating wind farms in different regions helps counteract the intermittent nature of wind power. Connections to hydro rich regions also allow for energy banking. In this regard, North American electric transmission infrastructure development could be viewed as a mechanism for some provinces and states to reach their goal of GHG emissions control.[34]

This means that linkage in electricity infrastructure is bilateral and, in the long run, aimed at harmonizing GHG policy goals and outcomes. It also recognizes the deep energy integration between North American countries, as explained in Chapter 4. This linkage can also be perceived as market expansion. In this respect, integration under NAFTA has been successful. Nonetheless, this electricity exchange goes beyond markets, and has started to standardize energy portfolios on a transregional scale.

Faced with the fact that no comprehensive climate-change legislation has been passed in the US, the Canadian approach does not take into account local types of economies and needs, and Mexican states are not allowed to participate fully in these initiatives, LGs responded by creating TGERs. These regional initiatives represent one strategy of LGs in North America to develop local policies for climate change. These regions also address the issue of how localities can contribute to producing global public goods.

Thus, although global public goods tend to be provided by global governance schemes, to date, localities are the bodies that contribute to this provision. Global public goods are provided to some extent, territorialized and divided according to layers of climate governance and levels of convergence, harmonization and linkage. This is not new. What is new is that leadership is coming from LGs' sphere of authority. LGs appear as the last frontier where climate change can be operational. It is interesting to note that, once climate governance falls within this frontier, some of these jurisdictions have granted authority to a collective and regional administration. This regionalization avoids NAFTA schemes and develops its own logic.

This logic is based on several topics. The specific geographical production of energy sources is vital to developing mitigation strategies. Some of the LGs are reducing their coal-based energy production and beginning to exploit their own renewable potential. In this way, offset projects also portray differentiated types of ecosystems, where LGs can profit from having forests or certain clean energy sources.

34 National Energy Board—Canada, 2009b:29.

Another topic is the LGs' kind of economy. This is inevitably related to the amount of energy each member uses. Thus the TGERs are successful in incorporating each sector in different ways in respect of GHG reduction. TGERs have also succeeded in allowing LGs to propose and sometimes distribute allowances. It is too early to tell, but RGGI's allowance auctions have left its members with a good amount of profit.

Another fact that shows that TGERs may be a real option for providing global public goods is their establishment of linking mechanisms with other levels of climate governance. Secondary markets in the three initiatives may perform as the means to involve global and national governance institutions in GHG carbon markets. Taking the concept of linking to the next level, TGERs are consolidating integration processes. This is exemplified by infrastructure integration in the electricity sector. Some facilities are now being integrated in cleaner processes due to their participation in TGERs. These regions have gone so far as to require exporters of electricity to achieve certain percentages of clean energy established in the region. Integration is also present in some of the developed regional policies. Some of them go to the next level: convergence. This element appears to be important in that it encompasses a transnational scope. In TGERs in North America, LGs from the three countries are either members or observers. In other words, their participation in coherent regions exemplifies a new way of local relating schemes beyond national ones.

If we compare the depth of the level of linkage in both TGERs, the deepest exchanges exist in the category of approaches and instruments. Shared ideas and operative mechanisms are the main connectors. Contrary to what some scholars may think, for the cases of transregional carbon markets in North America, the policy goals and outcomes have difficulty relating to LGs.[35]

This situation implies that the concept of transnational regionalization is indeed driving climate approaches, policies and infrastructure in the region. In this same way, it underscores the argument that integration in North America is taking place around a mid-range scope: regarding different levels of climate governance linking LGs with regional, national or global governance institutions, as well as establishing relations between TGERs themselves.[36] This integration has also been differentiated into sectors. Thus there is not just one comprehensive and formal regionalization process in North America, but several schemes with diverse rhythms and forms.

In spite of multi-directional efforts, the examples of linkage operating at the moment act unilaterally, and they raise several questions about their legality and legitimacy that will be discussed in the Conclusions.

---

35 For some authors, convergence in environmental issues is most pronounced at the policy level and less so at the instrumental level. For a detailed and contextualized explanation of this argument, see Wälti, 2010.

36 For further detail of the concepts of verticality and horizontality in policy integration, see Kern, 2010

Conclusions: TGERs and the Reconfiguration of Climate Governance in North America

The global governance approach has been based on the premise that, for global problems and provision of global public goods, global solutions and global institutions are required. The intention of this book has been to challenge that idea by arguing that regionalization processes with a transnational scope can better provide solutions to global problems such as climate change.

As proposed in Chapter 1, the problem of climate change can be analyzed under the concept of global public goods (or, in this case, "bads"). In their pure form, these goods are non-rival and non-exclusive. Regionalization of climate governance through TGERs suggests that public goods have partial exclusiveness and rivalry. The partial nature of the management of GHGs exemplifies a very practical approach in which emissions can be traced by pointing out their specific sources (upstream or downstream). It also tells us about the globalism of GHGs: once emitted, they affect everyone. In other words, the impact of specific-source GHG emissions does not recognize borders. The fact that TGERs are trying to deal with this partial provision is the expression of re-territorialization.

Alternative responses to address climate change have emerged in the context of the apparent failure of global climate governance. The KP has shown conflicting goals of formal and informal norms in several ways. First, it failed to achieve its fundamental goal, the reduction of GHGs and the consequent provision of clean air as a global public good. Second, it failed in creating incentive mechanisms for states to implement their commitments. Third, it failed in that it did not take into account the economic situation of the productive regions within countries, as well as the interests of companies and society. Fourth, it did not include potential polluters that at some point would act as free-riders, such as China or India (and other emerging economies). Fifth, the fact that two of the iconic leaders have withdrawn from the KP (the US and Canada) has left the whole governance scheme without legitimacy and a clear paymaster. Because of this, new forms of climate governance have appeared in order to counterbalance global failure.

TGERs in North America filled this gap through a different form of climate governance scheme. As presented in Chapter 1, the concept of governance can be understood in several ways. However, the proposal of this book has been to think of governance as processes of coordination among different actors and government levels in diverse constructed formal and informal relations expressed in multiple layers, with a leader granted legitimate authority. How have TGERs expressed this type of governance to address climate change?

As established in Chapter 4, TGERs are based on specific institutional architectures that allow them to do the task. To build on this institutional design it was necessary to redefine the concept of region and the structure and rules of TGERs. Rules are generally the coordination and operation processes of the general norms within these transregional initiatives. Coordination has gone from convergence of policies and harmonization to linkage, involving different degrees and forms of interconnection. In addition, these processes are not separate from the state and different levels of governance. To explain these relations, Chapter 1 established that climate governance has acted in the continuities of a sovereign state simultaneously with the discontinuities of a post-sovereign era. Continuities are represented in the analysis by the multi-level approach, which proposes different orders of government in addition to the state. To address discontinuities, a multi-layer approach was taken. A diversity of layers with networks of actors exemplifies a different relational process. The concept of TGERs aims to coordinate both approaches. However, this book has argued that the emphasis should be put on the leading actors in provision of global public goods, namely, local governments (LGs).

LGs have created institutional and formal connections among themselves to share rules and create (or reinforce) structures. Although this book has focused on the role of LGs, this approach to governance includes several levels of government, and at some point, business and society. TGERs started as informal institutions that have become formal ones through the signing of agreements and accords under the authority of LGs. In addition, there is a component of informal governance when speaking of the market approach used to address climate change and the reinforcement of norms. Both TGERs created carbon markets that are voluntary for each LG (as well as their commitments). In this way, informal agreements are enforced by the same organizational actors.[1] This formal–informal link opens the door to a multi-layer approach to climate governance in North America.

Thus LGs relocated authority by leading a multi-layer process with a transnational scope. There are several reasons for this relocation to take place in the case of climate change. The first is that historical relations between the national governments and LGs in North America have not been fully cooperative in this area. As explained in Chapter 2, the governments of Canada, the US and Mexico had tried to pursue national climate legislation and to include LGs. Mexico is the only country that has enacted a national climate-change law, but, as explained, this is starting to cause problems with LGs' prior climate actions: relations between the federal government and some states have been framed by competition. In the case of Canada, some years ago national legislation to enforce the country's commitments to the KP was passed. Nonetheless, provinces have not

1 Farrell and Héritier, 2003.

quite implemented them, and they have not been willing to absorb the economic costs imposed from Ottawa. In the US, LGs have taken the lead in establishing their own mechanisms and instruments, such as carbon markets.

The existence of TGERs reveals that a different form of governance is taking place in North America. LGs have formed new transnational institutions of climate governance that deal with the regionalization process in relation to climate and energy policy innovation. The Regional Greenhouse Gas Initiative (RGGI) and the Western Climate Initiative (WCI) are dynamic and flexible transregional institutions with the adaptive capacity to face change, especially resulting from potential national policies or global agreements. It is worth noting that transnational relations are more profound for US states than for Canadian provinces or Mexican states (with the exception of the US–Mexico neighboring states, which are deeply related in a bilateral way). Canadian provinces participated actively in the first stages of both initiatives; however, in the end, only Quebec remains as part of one of them. Mexican states have only been observers to these TGERs, although some of them are physically interconnected to them through the provision of electricity, relating them indirectly to climate transregional policies.

Apart from the creation of transnational institutions, the structure of both TGERs coincided at first with geographical neighboring spaces. This accounted for the natural ability of the space to use and produce certain kinds of fuels for their economies. It also accounted for the level of integration of trade and communication among them. However, a process of reinterpretation had to be developed for WCI, from which the majority of its partners withdrew in 2011. The only remaining members are California and Quebec. As explained in Chapter 6, this relation removes the geographical element from the regionalization process to achieve a more policy-oriented scheme of transregional cooperation.

Especially for RGGI, interconnection is evidenced by policy and also by infrastructure, which becomes an expression of harmonization and linkage of policies. In the case of TGERs, infrastructure had been connected before these were formed, especially in the energy sector. Oil and gas pipelines and electricity transmission lines have been present since the mid-twentieth century. As presented in Chapter 3, this integration was encouraged by NAFTA and by the NAEWG. In the context of NAFTA, the NAAEC appeared as an important regional effort through the Commission for Environmental Cooperation (CEC). This commission, along with the energy group of the NAEWG, has promoted the creation of information databases for decision-making. However, they have failed to lead the integration among both sectors and regional actors.

Under TGERs, these physical connections operated originally under multinational or binational agreements; they are now developed under regional rules to provide their services and products. Since TGERs appeared and linked energy provision/production to climate change, new energy mixes or portfolios have emerged locally. With the sharing and common understanding of certain environmental standards (which may converge, harmonize or link), this infrastructure is starting to change and adapt to renewable technologies and sources. In the same way, this regionalization has impacted social practices within the regions through instruments that directly include consumers, such as net metering systems or green markets.

Challenges and Limits to TGERs

In this optimistic scenario, RGGI and WCI have been operating their new carbon markets with relatively good results in institutionalization processes, GHG reduction, and income derived from the trade in emissions. There are opportunities to continue the development of this innovate form of climate governance. However, these transregional institutions face challenges and criticisms.

Challenges are present in several areas. The next section addresses limitations to this climate mechanism regarding the relation with federal governments' legislation and international agreements, the operation of the cap-and-trade system, its real impact on GHG reduction, and institutional robustness of carbon markets within TGERs. Of course, as well as these, there are power-game and political calculations of decision-makers at electoral times.

LGs Facing National Legislation and Global Governance

The most important limitations and challenges to the operation of TGERs to address climate change have to do with the approach they have chosen: carbon markets. How constitutional these markets are is at stake. As Richardson notes, states in the US are able to enact their own climate policies, including emission trading schemes (ETS). However, they violate certain pieces of commercial national legislation. A first example of this situation is the Dormant Commerce Clause, which prevents discrimination against out-of-state businesses, such as the coal industry in RGGI. In general, this clause serves to restrain states both from adopting protectionist policies (policies that disproportionately benefit in-state interests) and from adopting measures that result in excessive interference in foreign affairs.

As explained by several experts, under the Dormant Commerce Clause, interstate trading and offsetting programs could be characterized as a restriction on the movement of emissions-intensive products or services across state lines. Hence the response to concerns that RGGI's cap on electricity generators within the region will result in leakage; regulators are seeking ways to limit the importation of electricity by out-of-state generators not covered by RGGI's emissions cap. These "border adjustments" of RGGI are designed to encourage mitigation globally. However, this case may also violate some of the World Trade Organization's and different free trade agreements worldwide.[2]

2 Aldy and Stavins, 2011:7; LaMotte et al., 2009.

Another example of constitutional violation in the US is the Compact Clause, which forbids states to enter into an agreement or compact with other states or with a foreign power without the consent of Congress.[3] In 2011, the state of New York was challenged in court with the argument that the RGGI's Memorandum of Understanding violated the Compact Clause. The lawsuit (Thrun *v*. Cuomo, Index No. 4368-11 – New York Supreme Court Albany County, 2012) was filed by three state residents who argued that the program was effectively a coercive tax on electricity levied without the approval of the state legislature, in violation of the state constitution. The court found that the plaintiffs lacked standing to sue because they did not face an injury that could be distinguished from effects of the program on the general public. The court also found that the case was dismissed because of an "unreasonable delay" between 2008, when New York implemented its RGGI regulations, and 2011, when the law suit was introduced.[4]

Although the Compact Clause challenges are rare and have so far been unsuccessful, cap-and-trade schemes among states may well invoke renewed judicial application of this constitutional doctrine. However, as Richardson points out, in spite of these clauses, states will continue to use their carbon markets and linking systems.[5]

This situation rests on the argument that this type of decision has been left to the courts and, generally, they have favored the low-carbon policy trend of states and transregional initiatives. As La Motte and others note, even if an agreement among states, such as the RGGI, rises to the level of a constitutionally suspect "compact," courts have held that not all compacts require congressional consent. Instead, the key concern is whether the compact would encroach upon or interfere with the supremacy of the federal government.[6] Other federal legal instruments challenge the transregional initiatives in the US. Pre-emption is one of them. The doctrine of pre-emption derives from the Supremacy Clause, which states that federal laws are the supreme law of the land. Thus federal laws can pre-empt and invalidate state laws when (1) federal laws explicitly state that they pre-empt any state law addressing the subject of the federal legislation; when (2) Congress intends federal law to occupy the field; or when (3) a state law conflicts with a federal statute such that it is impossible for a private party to comply with both state and federal law or the state law stands as an obstacle to the accomplishment of Congress goals. As some experts note, the possibility to activate pre-emption will arrive as soon as a federal climate law is enacted.[7] However, some of the federal initiatives have considered this and designed some transition mechanisms, which have started by implementing some level of linkage to national agencies.

3 Richardson, 2012.

4 Georgetown Climate Center, 2013.

5 Richardson, 2012.

6 LaMotte et al., 2009. Another example of ruling in this sense is Rocky Mountain Farmers Union *v*. Goldstene (9th Cir. 2012) No. 12-15131.

7 LaMotte et al., 2009.

A final challenge to TGERs is the enactment of national policies with specific carbon-market instruments. If they were established and they created an obligatory carbon market, participants in regional initiatives would have to design adjustment mechanisms, especially for carbon prices. For example, "if a price is put to carbon, applying other policy tools such as renewable, energy efficiency or biofuel targets in addition to the carbon price can lead to overlap and might lock-in inefficient technologies."[8] Some analysts warn that, if national legislation in the US were to pass soon, it would outlaw the TGERs' efforts regarding the impact, scope and size of the markets.[9]

Problems with the Cap-and-trade System

Despite the "illegality" of carbon trading, some criticisms of the operation of the cap-and-trade system have been put on the table. For example, the Consilience Energy Advisory Group has highlighted that the main problems of carbon markets have to do more with the cap than the trading aspect: "Governments have failed to set sufficiently strict caps, so the price that has emerged from the market has been one that is too low to incentivize green behavior."[10] Thus Enion points out that caps need to be granted credibility by "de-politicizing" the system. He argues that political involvement diminishes legitimacy and the only way to avoid it is to provide access to equal and transparent information for stakeholders.[11]

In addition, the problem of leakage is present, especially with overlapping climate policy instruments where cherry-picking sometimes occurs. For example, emitters that comply with strict climate programs within their states and then join a new market with lower standards (another carbon market, local or foreign, a national scheme, or the flexible mechanisms of the KP) will over-comply. Therefore they will have excessive allowances or credits that can be traded to out-of-state emitters. The buyers of these credits can then emit the same amount of GHGs that the state sought to eliminate, with little environmental benefit.[12]

Overlapping problems and cherry-picking will occur if the observers to RGGI and WCI decide to participate in both schemes. Because both TGERs have different caps for markets and diverse requirements, these areas would need to solve the overlapping of expectations, priorities and rules.[13]

However, carbon markets are instruments in which cherry-picking takes place only in big corporations with different offices nationally and globally. In general, a local bidder can easily avoid leakage and will hardly over-comply targets. Some

8 Organization for Economic Cooperation and Development, 2009:13.

9 For a comparison between federal proposals and transregional initiatives, see Newcombe, 2008.

10 Consilience Energy Advisory Group Limited, 2009:219.

11 Enion, 2013.

12 Richardson, 2012; Linowes, 2010.

13 Cerny, 1999:190.

LGs set the bar too high for national compliance, and in the opinion of several scholars, in spite of the above-mentioned problems, this can be a positive incentive for national governments to standardize their policies at these levels.[14]

Nonetheless, LGs and transregional initiatives can also "race to the bottom" (pick the lowest standard available). This phenomenon has been studied in detail by Brown and Sovacol. These scholars note that the differences in scope and scale between carbon markets is discouraging, with RGGI addressing only CO_2 and one source (power plants), against WCI with all the GHGs recognized by the UNFCCC, or other schemes focusing on upstream or downstream fueling. They argue that so many different scopes and scales damage the functionality of an effective GHG global reduction. Indeed, for them, overlapping policies create uncertainty. For example, when integrating registries or GHG budgets, some local efforts overlap and become redundant. Industrial standards within a strict LG could cause loss of competitiveness and high costs when addressing different-level policies.[15]

In other words, as spillovers among participants start growing, the assumption that inter-jurisdictional competition benefits policies becomes questionable, because free-riders and negative externalities tend to appear.[16] As explained in Chapter 2, free-riding is a characteristic of the provision of global public goods. This phenomenon normally appears when some actors perform mitigating efforts and the rest do not, gaining benefits and absorbing no costs. This is a relation between the action of some versus the inaction of others (which, in general, are the majority). Thus RGGI and WCI would be victims of the rest of the US states, of both national governments, and of the rest of the world. Such a situation is common when national statistics present the aggregated GHG data that do not reflect the imbalance in efforts.

Another problem with local carbon markets is the management of sanctions, especially in mechanisms linking other actors outside the TGERs, such as offsets. The lack of real enforcement affects the functioning of the transregional carbon markets in North America.[17] Conflicts arise when enforcement agencies are responsible for carrying out their own policies and overlap with others. For example, the US EPA has its own law enforcement system and carries out its own inspections. On a national level, "implementation is likely swifter and more linear, but in a top down approach it can cause conflict, stalemate, or inaction."[18] Evidence of this argument is the California Independent System Operator, a "quasi-governmental agency" that regulates the reliability of the state's energy grid almost autonomously in the absence of national electricity regulation. A recent study warned that, as California tries to meet the stringent requirements of the Renewable Portfolio Standard of 33 percent renewable energy production,

14 Enion, 2013.

15 Brown and Sovacol, 2011.

16 Wälti, 2010:416.

17 Enion, 2013; Richardson, 2012.

18 Wälti, 2010:414.

so the need for flexible capacity resources arises. Thus this agency practically designs the rules to accomplish the goals and decides on how flexible it will be at certain times.[19]

In addition, there have been problems with other issues, such as whether the allowance allocation must be free or sold; if the price must have a floor and a ceiling; and how good the offsets and banking instruments are. Although these questions were individually addressed in the section regarding each market, it is necessary to note that, in carbon markets, participants (bidders) in RGGI and WCI rely on a cost–benefit analysis. This leads to the obvious point that carbon markets were not designed to manage GHG reduction, but to create new types of markets and give incentives to business to facilitate the integration of new energy sources. Thus the carbon markets' goals could be far from reducing GHGs and addressing environmental problems.[20]

This leads to a discussion of the range of TGERs. To date, RGGI deals only with CO_2 sources, and although WCI includes all GHGs, the Californian market has just started. It would be necessary, for convergence and harmonization purposes, to include the full range of GHGs and emission sources in both TGERs. This situation could avoid mitigation costs facing national policies in the long run. In a scenario where climate change would be addressed by controlling only CO_2, "the costs in 2050 would amount to 7% of the world GDP rather than 4% of world GDP if other gases were included."[21] However, one of the main criticisms of TGERs regards the real impact of addressing GHGs on a small scale. Both transregional initiatives are relatively new and the amounts of GHGs to be reduced are not very significant if compared to other large emitters such as China. For example, CO_2 reduction by RGGI represents only 10 percent of US GHG emissions.[22]

Therefore the question regarding the environmental impact of TGERs is one of the most important ones, and must be addressed soon. Contrary to the scientific debates on climate global governance, TGERs have missed the chance to include the scientific community in the design of carbon markets. In establishing the cap they have not taken into account the natural absorption rate of GHGs of their territories. This was indeed the basis of global climate governance, where climate scenarios designed by the IPCC were the starting point to establish reduction commitments and maximum temperature.[23] TGERs' initiatives are based on economic factors and the risks that local economies can afford to take, not on a reduction based on parts per million of GHGs in the atmosphere.

19 Grimes, 2012.

20 Arnaud et al., 2009.

21 Organization for Economic Cooperation and Development, 2009:1.

22 World Resources Institute, 2010.

23 Quadri, 2007.

Institutional Robustness and Political Legitimacy Put in Question

One of the main problems of transregional governance schemes has been institutional robustness. As explained in previous chapters, the institutional architecture needs to consolidate its goals and members. RGGI has not suffered a massive withdrawal, and, after the good results of recent years, it is expected that it will achieve consolidation. However, in WCI, it is quite symptomatic that the majority of the members withdrew just after the economic crisis emerged. Reasons were diverse. Some were genuinely owing to an economic cost–benefit analysis, but others were mainly due to political agendas. The remaining partners were those that do not rely on fossils fuels to move their economies: Quebec and California. As pointed out by some scholars, "the mortality rate for domestic bureaucracies is estimated to be five times lower than for international governmental organizations."[24]

In these situations the paymasters play a key role in holding together the initiative. Sonja Wälti mentions, in the context of the US, the "California effect," whereby the state has become a center for high-tech and low-pollution industries to suit the needs and skills of its pollution-sensitive population. She continues by arguing that the allocation of power, resources and organizational means across levels of government depends on institutional differences. In other words, the main determinant of robustness is how the institutional architecture develops through time and how well it links with other potential policy substitutes (as national policy or a new version of the KP).[25]

Institutional robustness is often sustained by political legitimacy granted by the use of profit from carbon markets for social ends. One of the main themes that can de-legitimize TGERs' approach is that LGs sometimes use cap-and-trade revenue to compensate for other government expenses and not for the purpose intended. For example, the government of New Hampshire, through its Legislature, took 3.1 million USD in RGGI funds to help balance the state budget for June 2010.[26] As a region, RGGI raised 792 million USD for clean energy programs; however, 25 percent of this revenue was used to complement local budgets.[27] There are other examples of this situation. As Hiskes mentions, in Delaware, 15 percent of the auction revenue helped low-income consumers seal their homes through a weatherizing assistance program and in Maryland funds were used to fund a weatherization workforce training program at 13 community colleges.[28] Although related to climate-changing conditions, these actions were taken by some constituents not as environmental ends, but as social policy already

24 Hooghe and Marks, 2003:239.

25 Wälti, 2010.

26 New Hampshire Business Resources, 2010.

27 Hiskes, 2010.

28 Ibid.

being financed by taxes. In the eyes of the public, this can harm the legitimacy of transregional efforts.

Although this book is not focused on the politics of TGERs in North America, it is worth noting that legitimacy and robustness depend, also, on local politics. In some states, alternate party governments have taken off the state's agenda the issue of climate change and their participation in transregional initiatives. Evidently, electoral processes always interfere with policy, as in the case of the Midwestern Greenhouse Gas Accord (MGGRA). In the RGGI and WCI regions, non-conservative parties' victories tend to keep the regional initiatives stable.[29]

Opportunities for TGERs and Optimistic Scenarios

In spite of these problems, transregional approaches in North America have clear advantages. The policy choices RGGI and WCI have facilitated experimentation and innovation. Because of the choice of carbon markets as the main tool, direct contact between companies and authorities is easier and quicker, so innovation tends to speed up as the correct incentives frame and encourage research. Closeness to authorities involved in regulation and GHG administration also facilitates accountability.[30] The fact that carbon markets in North America are administered by LGs' platform of government gives assurance to citizens in claiming any error in policy, or taking cases to court. In the end, the cap-and-trade system performs within a market approach, but where LGs are responsible for designing and enforcing public policy.

Some scholars have called this "climate capitalism," defining it as a model that squares capitalism's need for continual economic growth with substantial shifts away from carbon-based industrial development. Carbon markets are the first steps towards this in an era of unavoidable capitalism.[31] This is the case of RGGI and WCI. In the words of Robert Stavins, the RGGI can be considered "a great environmental success achieved by the 'fortunate coincidence' of low natural gas prices, economic recession, and mild weather. This is hardly something to be lamented."[32] Coincidence or not, other schemes, such as the three flexible

29 The case of California is *sui generis*. In general, either Republicans, as with their former governor, Arnold Schwarzenegger, or Democrats have supported environmental policies and the creation of WCI. However, generally in other US states, Democrats were the leaders of the original WCI. In the case of the Canadian provinces, liberal or left parties (Neodemocratic Party and Parti Québécois) have been leading transregional cooperation. See Hiskes, 2010; Gardner and Henderson, 2010. For a full report on each of the RGGI's electoral processes regarding the initiative, see Johnson, 2010.

30 Brown and Sovacol, 2011.

31 Newell and Paterson, 2010.

32 Stavins, 2012.

mechanisms of the KP or carbon taxes, have experienced the same fortunate conditions and their results have not been as expected.

TGERs also represent the possibility of serving as building blocks in a "middle-and-around" strategy to build larger carbon markets in North America, because they give credibility and are proven formulas. In other words, transregional initiatives are becoming the cornerstones in tackling climate change in North America, with small-scale, but real, results.

Although TGERs face the challenge of stability, one of the opportunities to keep the regions together is the creation of strong institutional schemes. For example, RGGI and WCI have created strong technical institutions to develop standards and instruments that allow their carbon markets to work.

The flexibility of RGGI and WCI opens the possibility for LGs to participate in national and international forums through different associations. LGs have presence in UNFCCC COPs through the International Council for Local Environmental Initiatives or other associations;[33] they also participate directly as a counterpart to their national governments, as some of the Canadian provinces have done.

RGGI and WCI recognize that global climate governance has forgotten several issues when designing instruments such as the KP. One of them is the fact that responsibility must be differentiated regionally and depend on local economic dynamics, as well as on the level of integration. This condition appears necessary to competitiveness, especially in regional mechanisms such as NAFTA.

In spite of unresolved issues, RGGI and WCI are demonstrating that a transregional approach is useful in reconfiguring climate governance in North America under the leadership of local governments.

33 International Council for Local Environmental Initiatives, 2010d.

Appendix A
Specific Commitments under the Kyoto Protocol

Annex I Parties	Commitments (percentage of reduction regarding base year)	CO_2e emissions in 1990 (1000 tons)	CO_2e emissions in 1990 (world percentage)
United States	93	4,957,022	36.1
Japan	94	1,173,360	8.5
Europe:	92	3,326,494	24.2
Germany	92	1,012,443	7.4
UK	92	584,078	4.3
Italy	92	428,941	3.1
France	92	366,536	2.7
Spain	92	260,654	1.9
Netherlands	92	167,600	1.2
Belgium	92	113,405	0.8
Greece	92	82,100	0.6
Sweden	92	61,256	0.4
Austria	92	59,200	0.4
Finland	92	53,900	0.4
Denmark	92	52,100	0.4
Portugal	92	42,148	0.3
Ireland	92	30,719	0.2
Luxemburg	92	11,343	0.1
Monaco	92	71	0
Canada	94	457,441	3.3
Australia	108	288,965	2.1
New Zealand	100	25,530	0.2
Switzerland	92	43,600	0.3

Norway	101	35,533	0.3
Iceland	110	2,172	0
Liechtenstein	92	208	0
Former Soviet Union:		3,417,981	24.9
Russia	100	2,388,720	17.4
Poland (1988)	94	414,930	3.0
Romania (1989)	92	171,163	1.2
Czech Republic	92	169,514	1.2
Bulgaria (1988)	92	82,990	0.6
Hungary (1985–87)	94	71,673	0.5
Slovakia	92	58,278	0.4
Estonia	92	37,797	0.3
Latvia	92	22,976	0.2
Croatia (no base year)	95	No data	No data
Lithuania	92	No data	No data
Slovenia (no base year)	92	No data	No data
Ukraine (no base year)	100	No data	No data
Total		13,728,306	

Source: Bartsch and Müller, 2000:17; United Nations Framework Convention on Climate Change, 1998.

Appendix B: Chronology of COP Meetings

COP 1 1995 Berlin, Germany	The Parties expressed concerns about countries' abilities to meet commitments under the Convention. These were expressed in a UN ministerial declaration known as the "Berlin Mandate," which established a two-year Analytical and Assessment Phase (AAP) to negotiate a "comprehensive menu of actions" for countries to pick from and choose future options to address climate change that for them, individually, made the best economic and environmental sense. The Berlin Mandate exempted non-Annex I countries from additional binding obligations, in keeping with the principle of "common but differentiated responsibilities" established in the UNFCCC – even though, collectively, the larger, newly industrializing countries were expected to be the world's largest emitters of greenhouse gas emissions 15 years hence.
COP 2 1996 Geneva, Switzerland	A Ministerial Declaration was adopted on July 18, and reflected a U.S. position statement presented by Timothy Wirth, former Under Secretary for Global Affairs for the U.S. State Department at that meeting, which accepted the scientific findings on climate change proffered by the Intergovernmental Panel on Climate Change (IPCC) in its second assessment (1995); rejected uniform "harmonized policies" in favor of flexibility; and called for "legally binding mid-term targets."
COP 3 1997 Kyoto, Japan	The Kyoto Protocol was adopted. Most industrialized nations and some central European economies in transition agreed to legally binding reductions in greenhouse gas emissions of an average of 6 to 8 percent below 1990 levels between the years 2008 and 2012, defined as the first emissions budget period. The United States would be required to reduce its total emissions an average of 7 percent below 1990 levels. However, neither the Clinton administration nor the Bush administration sent the protocol to Congress for ratification. The Bush administration explicitly rejected the protocol in 2001.

COP 4 1998 Buenos Aires, Argentina	It had been expected that the remaining unresolved issues in Kyoto would be negotiated at this meeting. Due to the complexity of these issues, the parties could only adopt a two-year "Plan of Action" to advance efforts and to devise mechanisms for implementing the Kyoto Protocol, to be completed by 2000.
COP 5 1999 Bonn, Germany	Ministers and officials designed a timetable for completing unresolved details of the Kyoto Protocol by COP-6 from November 13–24, 2000 in The Hague. It was primarily a technical meeting, and did not reach major conclusions.
COP 6 2000 The Hague, Netherlands	No agreement was reached. Formal Plenary decides to suspend COP-6 and reconvene in 2001.
COP 6 Reconvened 2001 Bonn, Germany	United States withdrew from the Kyoto Protocol. As the other parties negotiated the key issues, agreement was reached on most of the major political issues, to the surprise of most observers given the low level of expectations that preceded the meeting. The agreements included: Flexible Mechanisms that the United States had strongly favored as the Protocol was initially put together, including emissions trading; Joint Implementation (JI); and the Clean Development Mechanism (CDM), which allows industrialized countries to fund emissions reduction activities in developing countries as an alternative to domestic emission reductions.
COP 7 2001 Marrakesh, Morocco	Negotiators completed the Buenos Aires Plan of Action, finalizing most of the operational details and setting the stage for nations to ratify the Protocol. The completed package of decisions is known as the "Marrakesh Accords."
COP 8 2002 New Delhi, India	It was acknowledged that vulnerable groups, such as children, would have to be protected. Poverty was also addressed as one of the main problems of developing countries; it was said to be an obstacle for mitigating climate change and a reason for difficulty in adapting.
COP 9 2003 Milan, Italy	The Parties discussed adaptation, mitigation and technological advance, as well as the need for strengthening implementation by committed states through institution building.

COP 10 2004 Buenos Aires, Argentina	This meeting marked the 10th anniversary of the entry into force of the UNFCCC, which served as a central theme for the meeting. The accomplishments of the past 10 years were revised and challenges such as adaptation and mitigation policies were discussed, as well as technology.
COP 11 2005 Montreal, Canada	The Kyoto Protocol entered into force. COP 11 was also the first meeting of the parties to the Kyoto Protocol (CMP 1). It was therefore one of the largest intergovernmental conferences on climate change ever. The Montreal Action Plan pledged for extending the life of the Kyoto Protocol beyond its 2012 expiration date and negotiating deeper cuts in greenhouse-gas emissions. The conference attracted attention from business as a result of the starting operations of the European emissions trading scheme and the Clean Development Mechanism, a tool to promote sustainable development and combat climate change.
COP 12 2006 Nairobi, Kenya	China, India, South Africa, Brazil and Mexico propose the creation of financial mechanisms through the Global Fund.
COP 13 2007 Bali, Indonesia	The Parties discussed and enacted the Bali Plan, designing a calendar for preparing for a new agreement after the end of the KP in 2012.
COP 14 2008 Poznan, Alemania	It was established that industrialized countries would pay 2 percent for financing Credits for Emissions Reductions at the Clean Development Mechanism.
COP 15 2009 Copenhagen, Denmark	It was agreed that the global temperature could not go beyond 2 °C, which would stabilize CO_2e at 450ppm.
COP 16 2010 Cancun, Mexico	The Cancun Accords were passed by all partners, except for Bolivia. These accords establish financial schemes for implementing mitigation and adaptation, especially in the developing world. In addition, the REDD mechanisms (reduction of emissions from deforestation or degradation) were recognized as sources for mitigation.
COP 17 2011 Durban, South Africa	Scientific assessments noted that the 2 °C specified was not enough and that the goal would need to be at least 3.5 °C. Canada withdrew from the KP and China and India refused to commit to specific GHG reductions.

COP 18 2012 Doha, Qatar	The KP was extended to 2020 and a calendar towards a new agreement was set.

Source: Author's design with data from United Nations Framework Convention on Climate Change, 2013b and 2011f; and Chinaview, 2007.

Bibliography And Information Sources

Abraham, J. (2009) Renewable Energy Supply and Transmission (Tucson, AZ: Southwest Climate Change Network) at http://www.southwestclimatechange.org/solutions/reducing-emissions/energy-supply-transmission (cited: 06/15/11).

Adams, W.M. (2009) Green Development: Environment and Sustainability in a Developing World (New York: Routledge).

Agrawala, Sh. and A. Steinar (2001) The Two-level Games and the Future of Climate Changes, Energy & Environment, Vol. 12, Nos. 2–3, pp. v–xi (Lysaker, Norway: The Fridtjof Nansen Institute).

Aguayo, F. (2010) Climate Change Mitigation and Institutions for Sustainable Development? (Mexico City: El Colegio de México).

Aguirre, I. (1999) Making Sense of Paradiplomacy? In F. Aldecoa and M. Keating (eds.) Paradiplomacy in Action: The Foreign Relations of Subnational Governments (London: Frank Cass Publishers).

Aldecoa, F. and M. Keating (eds.) (1999) Paradiplomacy in Action: The Foreign Relations of Subnational Governments (London: Frank Cass Publishers).

Aldy, J.E. and R.N. Stavins (2011) The Promise and Problems of Pricing Carbon: Theory and Experience, Working Paper No. 17569 (Cambridge, MA: National Bureau of Economic Research) at http://www.nber.org/papers/w17569 (cited: 06/17/13).

Aldy, J.E. and R.N. Stavins (eds.) (2007) Architectures for Agreement: Addressing Global Climate Change in the Post-Kyoto World (Cambridge, UK: Cambridge University Press).

Alfie, M. (2009) North American & European Environmental Agencies: Comparative Perspectives. In I. Hussain (ed.), North America at the Crossroads: NAFTA after 15 Years (Mexico City: Universidad Iberoamericana, A.C.).

Amen, M., N.J. Toly, P.L. McCarney and K. Segbers (eds.) (2011) Cities and Global Governance: New Sites for International Relations (Farnham, UK: Ashgate Publishing).

Andrews Kurth Attorneys (2010) Andrews Kurth Climate Change Outline (The Woodlands, TX: Andrews Kurth Attorneys) at http://www.andrewskurth.com/assets/htmldocuments/AKClimateChange16.pdf (cited: 07/02/13).

Antal, E. (2004) Cambio climático: Desacuerdo entre Estados Unidos y Europa (Mexico: UNAM/Plaza y Valdés).

Arnaud, B., E. Nick and N. Howarth (2009) Carbon Markets: An International Business Guide (London: Earthscan).

Arrhenius, S. (1896) On the Influence of Carbonic Acid in the Air upon the Temperature of the Ground, Philosophical Magazine and Journal of Science, Series 5, Vol. 41 (April), pp. 237–276 (London: Taylor & Francis).

Ayres, J. and L. Macdonald (eds.) (2009) Contentious Politics in North America: National Protest and Transnational Collaboration under Continental Integration (London: Palgrave Macmillan).

Bailey, R.G. (1998) Ecoregions: The Ecosystem Geography of the Oceans and Continents (London: Springer).

Bakvis, H. and G. Skogstad (2002) Canadian Federalism: Performance, Effectiveness, and Legitimacy (Don Mills, ON: Oxford University Press).

Baldersheim, H. and M. Ogard (eds.) (2011) The Rise of the Networking Region: The Challenges of Regional Collaboration in a Globalized World. In H. Baldersheim, V.H. Are and M. Ogard (eds.) The Rise of the Network Region: The Challenges of Regional Collaboration in a Globalized World (Farnham, UK: Ashgate Publishing).

Baldersheim, H., V.H. Are and M. Ogard (eds.) (2011) The Rise of the Network Region: The Challenges of Regional Collaboration in a Globalized World (Farnham, UK: Ashgate Publishing).

Balibar, E. (2004) Europe as a Borderland. In The Alexander Von Humboldt Lecture in Human Geography (Netherlands: University of Nijmegen).

Banting, H. and R. Simeon (eds.) (2006) Degrees of Freedom: Canada and the United States in a Changing World (Montreal, QC and Kingston, ON: McGill-Queen's University Press).

Barnes, J. (2011) Oil and US–Mexico Bilateral Relations. In Baker Institute Energy Forum, the Future of Oil in Mexico (Mexico City: James A. Baker III Institute for Public Policy and Oxford University).

Bartsch, U. and B. Müller (2000) Fossil Fuel in a Changing Climate: Impacts of the Kyoto Protocol and Developing Country Participation (Oxford: Oxford University Press).

Battarbee, R.W. and H.A. Binney (2008) Natural Climate Variability and Global Warming (London: Wiley-Blackwell).

Baumol, W.J. and W.E. Oates (1988) The Theory of Environmental Policy, Second Edition (Cambridge, UK: Cambridge University Press).

BBC News (2008) California Sues US over Emissions (London: BBC News) at http://news.bbc.co.uk/2/hi/americas/7169200.stm (cited 06/15/10).

Bernstein, S., J. Brunnée, D. Duff and A. Green (eds.) (2008) A Globally Integrated Climate Policy for Canada (Toronto, ON: University of Toronto Press).

Bertrand, C., J.P. van Ypersele and A. Berger (2002) Are Natural Climate Forcings Able to Counteract the Projected Anthropogenic Global Warming?, Climatic Change, Vol. 55, pp. 413–427 (Netherlands: Kluwer Academic Publishers).

Betsill, M. (2006) NAFTA as a Forum for CO_2 Permit Trading. In H. Selin and S. Vandeveer (eds.) Climate Change Politics in North America: The State of Play (Washington, DC: Woodrow Wilson International Center for Scholars and

Canada Institute) at http://www.wilsoncenter.org/events/docs/CI_OccPaper_ClimateChange.pdf (cited: 05/05/09).

Biersteker, Th. J. and R.B. Hall (2002) Private Authority as Global Governance. In Th. J. Biersteker and R.B. Hall (eds.) The Emergence of Private Authority in Global Governance (Cambridge, UK: University of Cambridge Press).

Boas, M., M.H. Marchand and T.M. Shaw (2005) The Political Economy of Regions and Regionalisms: An Introduction to Our Critical, Revisionist Inquires. In M. Boas, M.H. Marchand and T.M. Shaw (eds.) The Political Economy of Regions and Regionalisms (Basingstoke: Palgrave Macmillan).

Bodansky, D.M. (1999) The Legitimacy of International Governance: A Coming Challenge for International Environmental Law?, Scholarly Works, Paper 443 (Athens, GA: University of Georgia School of Law) at http://digitalcommons.law.uga.edu/fac_artchop_443 (cited: 05/20/11).

Boesen, N. (2006) Governance and Accountability—How Do Formal and Informal Interplay and Change?, International Seminar: Informal Institutions and Development (Paris: OECD) at http://www.oecd.org/dataoecd/52/38/37790183.pdf (cited: 03/18/10).

Börzel, T. and Th. Rise (2002) Public–Private Partnerships: Effective and Legitimate Tools of International Governance? In E. Grande and L.W. Pauly (eds.) Complex Sovereignty: On the Reconstitution of Political Authority in the 21st Century (Toronto, ON: University of Toronto Press).

Bothwell, R. and J. Daudelin (eds.) (2008) Canada among Nations: 100 Years of Canadian Foreign Policy (Montreal, QC and Kingston, ON: McGill-Queen's University Press).

Bourse de Montréal (2009) Database (Montreal, QC: Bourse de Montréal) at http://www.m-x.ca/accueil_fr.php?changeLang=yes& (cited: 01/28/09)

Boyer, M.A. (2013) Global Climate Change and Local Action: Understanding the Connecticut Policy Trajectory, International Studies Perspectives, Vol. 14, No. 1 (February), pp. 79–107 (Malden, MA: Wiley Periodicals, Inc.).

Bradley, R.S. (2000) Past Global Changes and Their Significance for the Future, Quaternary Science Reviews, Vol. 19, Nos. 1–5, pp. 391–402 (London: Elsevier Ltd) at http://www.sciencedirect.com/science/article/pii/S0277379199000712 (cited: 03/26/11).

Bramley, M. (2000) A Climate Change Resource Book for Journalists (Ottawa, ON: Pembina Institute and Government of Canada) at http://www.pembina.org/pub/25 (cited: 04/05/08).

British Columbia Hydro Regulatory Group (2009) Electricity Rate Comparison Annual Report (Vancouver: British Columbia Hydro Regulatory Group) at http://www.bchydro.com/etc/medialib/internet/documents/appcontent/your_account/na_rates_compare.Par.0001.File.na_rates_compare.pdf (cited 05/12/10).

British Columbia Ministry of Environment (2010) BC Holds First Carbon-neutral WCI Meeting (Vancouver: British Columbia Ministry of Environment) at http://www2.news.gov.bc.ca/news_releases_2009-2013/2010ENV0006-000223.htm (cited: 02/02/11).

Brousseau, E., T. Dedeurwaerde and B. Siebenhüner (2012) Reflexive Governance for Global Public Goods (Cambridge, MA: Massachusetts Institute of Technology).

Brovkin, V., S. Sitch, W. von Bloh, M. Claussen, E. Bauer and W. Cramer (2004) Role of Land Cover Changes for Atmospheric CO_2 Increase and Climate Change during the Last 150 Years, Global Change Biology, Vol. 10, pp. 1–14 (Hoboken, NJ: Wiley-Blackwell).

Brown, D. (2002) Aspects of Multilevel Governance in Australia and Canada (Kingston, ON: Queen's University) at http://www.pinkcandyproductions.com/portfolio/conferences/globalization/pdfs/brown.pdf (cited: 04/14/11).

Brown, D., S. Khagram, M. Moore and P. Frumkin (2000) Globalization, NGOs, and Multisectoral Relations. In J.S. Nye and J. Donahue (eds.) Governance in a Globalizing World (Washington, DC: Brookings Institution Press).

Brown, M.A. and B.K. Sovacol (2011) Climate Change and Global Energy Security: Technology and Policy Options (Cambridge, MA: Massachusetts Institute of Technology).

Brunet-Jailly, E. (2011) The Emergence of Cross-border Regions and Canadian–United States Relations. In M. Amen, N.J. Toly, P.L. McCarney and K. Segbers (eds.) Cities and Global Governance: New Sites for International Relations (Farnham, UK: Ashgate Publishing Inc.).

Bryner, G.C. (2004) Global Interdependence. In R. Durant, D. Fiorino and R. O'Leary (eds.) Environmental Governance Reconsidered: Challenges, Choices and Opportunities (Cambridge, MA: MIT Press).

Buchdahl, J. (2002) Climate Change: Atmosphere (Manchester: Climate and Environment Information Programme and Manchester Metropolitan University).

Bureau of the Census—US (2013) US and World Population Clock (Washington, DC: United States Census Bureau) at http://www.census.gov/popclock/ (cited: 07/09/13).

Bureau of the Census—US (2009) US Population by State, 1790 to 2009 (Upper Saddle River, NJ: Pearson Education) at http://www.infoplease.com/ipa/A0004986.html#axzz0yKNHetHD (cited: 03/28/11).

Burton, I., E. Diringer and J. Smith (2006) Adaptation to Climate Change: International Policy Options (Arlington, VA: Pew Center on Global Climate Change) at http://www.pewclimate.org/docUploads/PEW_Adaptation.pdf (cited: 05/03/11).

Busby, J. (2010) After Copenhagen: Climate Governance and the Road Ahead, Working Paper (New York: Council on Foreign Relations, Inc.).

Buzan, B. and R. Little (2000) International Relations Theory (Oxford: Oxford University Press).

Caldwell, S. (2009) Cap-and-trade Modeling: Initial Policy Run Results (Miami, FL: Pew Center on Global Climate Change) at http://www.midwesternaccord.org/Meeting%2520material%2520pages/GHG-meeting-9/Modeling_Cap-and-Trade_Results_FINAL.pdf (cited: 02/11/11).

California Air Resources Board (2013a) California Air Resources Board Quarterly Auction 1, November 2012—Summary Results Report (Sacramento, CA: California Air Resources Board & Western Climate Initiative, Inc.) at http://www.arb.ca.gov/cc/capandtrade/auction/auction.htm (cited: 07/01/13).

California Air Resources Board (2013b) Compliance Instrument Tracking System Service (Sacramento, CA: California Air Resources Board) at http://www.arb.ca.gov/cc/capandtrade/markettrackingsystem/markettrackingsystem.htm (cited: 07/01/13).

Cámara de Senadores (2011a) Boletín 0447 Aprueba el Senado expedir Ley General de Cambio Climático (Mexico City: Cámara de Senadores) at http://comunicacion.senado.gob.mx/index.php?option=com_content&view=article&id=2743:boletin-0447-aprueba-senado-expedir-ley-general-de-cambio-climatico&catid=46:boletin-de-prensa&Itemid=177 (cited: 12/17/11).

Cámara de Senadores (2011b) Dictamen de la Ley General de Cambio Climático (Mexico City: Cámara de Senadores).

Cammack, R., T. Atwood, P. Campbell, H. Parish, A. Smith, F. Vella and J. Stirling (2008) Oxford Dictionary of Biochemistry and Molecular Biology, Second Edition (Oxford: Oxford University Press).

Campbell, A. 2009. Science in Environmental Policy: The Politics of Objective Advice (Cambridge, MA: MIT Press).

Canadian Centre for Energy Information (2011) Province Profile by the Numbers (Calgary, AB: Canadian Centre for Energy Information) at http://www.centreforenergy.com/Documents/AboutEnergy/ByTheNumbers/BC-bythenumbers.pdf (cited: 05/17/13).

Canadian Electricity Association (2012) Key Canadian Electricity Statistics (Ottawa, ON: Canadian Electricity Association) at http://www.electricity.ca/media/Industry%20Data%20and%20Electricity%20101%20May%202012/KeyCanadianElectricityStatistics_2012.pdf (cited: 07/01/13).

Canadian Electricity Association (2011) Reliability: Ensuring Reliability through NERC Standards (Ottawa, ON: Canadian Electricity Association) at http://www.electricity.ca/industry-issues/economic/reliability.php (cited 02/13/11).

Canadian Electricity Association (2009) Enhancing Our Integrated Electricity System: An Opportunity to Build on Success (Ottawa, ON: Canadian Electricity Association) at www.canelect.ca (cited 02/13/11).

Canadian Electricity Association (2006) Power Generation in Canada, A Guide (Ottawa, ON: Canadian Electricity Association) at http://www.canelect.ca/en/PDFs/HandBook.pdf (cited 02/13/11).

Capello, R., P. Nijkamp and G. Pepping (1999) Sustainable Cities and Energy Policies (Berlin: Springer-Verlag).

Cárdenas, A. (2010) Ley General sobre Cambio Climático (Mexico City: Senado de la República) at www.senado.gob.mx/iilsen/content/lineas/docs/cambio_climatico_lxi/iniciativa%20243.pdf (cited: 17/12/10).

Catenacci, M. (2010) Environmental Policy Integration at the Regional and Local Levels, Setting the Problem into Context. In A. Goria, A. Sgobbi and I. von Homeyer (eds.) Governance for the Environment: A Comparative Analysis of Environmental Policy Integration (Cheltenham, UK: Edward Elgar Publishing).

Cerny, Ph. (2006) Political Globalization and the Competition State. In H. Banting and R. Simeon (eds.) Degrees of Freedom: Canada and the United States in a Changing World (Montreal, QC and Kingston, ON: McGill-Queen's University Press).

Cerny, Ph. (1999) Globalization, Governance, and Complexity. In J. Hart and A. Pakrash (eds.) (1999) Globalization and Governance (London: Routledge).

Chacón, D., M.E. Giner, M. Vázquez-Valles, S. Roe, J.A. Maldonado, H. Lindquist, B. Strode, R. Anderson, C. Quiroz and J. Schreiber (2010) Emisiones de Gases de Efecto Invernadero en Chihuahua y Proyecciones de Casos de Referencia 1990-2025 (Cd. Juárez: Border Environment Cooperation Commission, Center for Climate Strategies, Government of Chihuahua).

Chicago Climate Exchange (2009) Database (Chicago, IL: Chicago Climate Exchange) at http://www.chicagoclimatex.com/ (cited: 01/28/09).

Chinaview (2007) Backgrounder: Chronology of UNFCCC (Beijing: Xinhua News Agency) at www.chinaview.cn (cited: 06/30/11).

Clark, T. (2002) The Policy Process: A Practical Guide for Natural Resources Professionals (New Haven, CT: Yale University Press).

Clarke, T. (2008) Tar Sands Showdown: Canada and the New Politics of Oil in an Age of Climate Change (Toronto, ON: James Lorimer & Company, Ltd.).

Clarkson, S. (2010) Multi-modal Governance in North America. In H. Enderlein, S. Wälti and M. Zürn (eds.) Handbook on Multi-level Governance (Cheltenham, UK: Edward Elgar Publishing).

Clarkson, S. (2008) Does North America Exist? Governing the Continent after NAFTA and 9/11 (Toronto, ON: University of Toronto Press).

Clarkson, S. and M. Mildenberger (2011) Dependent America? How Canada and Mexico Construct US Power (Toronto, ON: University of Toronto Press).

Coase, R. (1993) The Problem of Social Cost. In R. Dorfman and N. Dorfman (eds.) Economics of the Environment, Selected Readings, Third Edition (New York: W.W. Norton and Co.).

Cohen, S., B. Bass, D. Etkin, B. Jones, J. Lacroix, B. Mills, D. Scott and C. Van Kooten (2004) Regional Adaptation Strategies. In H. Coward and A. Weaver (eds.) Hard Choices: Climate Change in Canada (Waterloo, ON: Wilfrid Laurier University Press).

Collier, D. and J.E. Mahon Jr. (1993) Conceptual "Stretching" Revisited: Adapting Categories in Comparative Analysis, American Political Science Review, Vol. 87, No. 4, pp. 845–855 (Cambridge, UK: Cambridge University Press).

Collier, R.B. and D. Collier (1991) Shaping the Political Arena (Princeton, NJ: Princeton University Press).

Comisión de Cooperación Ecológica Fronteriza (2010) Proyectos certificados (Cd. Juárez: Comisión de Cooperación Ecológica Fronteriza) at http://www.cocef.org/espanol/index.html (cited: 01/22/11).

Commission for Environmental Cooperation (2013) Climate Change: Low-carbon Economy (Montreal, QC: Commission of Environmental Cooperation) at www.cec.org/Page.asp?PageID=1323&SiteNodeID=1229 (cited: 06/10/13).

Commission for Environmental Cooperation (2012) Environmental Atlas of North America (Montreal, QC: Commission of Environmental Cooperation) at http://www.cec.org/naatlas (cited: 05/27/2013).

Commission for Environmental Cooperation (2009) The Three Countries Working Together to Protect Our Shared Environment (Montreal, QC: Commission of Environmental Cooperation) at http://www.cec.org/PageID=1226&SiteNodeID=310&BL_ExpandID=154 (cited: 06/10/13).

Commission for Environmental Cooperation (2003) Total Installed Capacity of Renewables in North America (2003) (Montreal, QC: Commission of Environmental Cooperation) at http://www.cec.org/Page.asp?PageID=122&ContentID=1330&SiteNodeID=500&BL_ExpandID (cited: 10/27/2008).

Commission for Environmental Cooperation (1993) North American Agreement on Environmental Cooperation (Montreal, QC: Commission of Environmental Cooperation) at http://www.cec.org/pubs_info_resources/law_treat_agree/naaec/index.cfm?varlan=english (cited: 10/27/2008).

Commission on Global Governance (1995) Our Global Neighborhood (New York: Oxford University Press).

Conca, K. (2005) Governing Water: Contentious Transnational Politics and Global Institution Building (Cambridge, MA: MIT Press).

Consilience Energy Advisory Group Limited (2009) Climate Change & Emissions Trading: What Every Business Needs to Know, Third Edition (London: Consilience Energy Advisory Group Limited).

Cooper, A. and D. Rowlands (2006) Canada among Nations 2006 (Montreal, QC and Kingston, ON: McGill-Queen's University Press).

CRC Magazine (2009) Carbon Reduction Commitment Allowances (London: CRC Magazine) at http://www.carbonreductioncommitment.info (cited: 06/18/11).

Crowley, Th. J. (2000) Causes of Climate Change over the Past 1000 Years, Science, Vol. 289, No. 5477, pp. 270–277 (Washington, DC: American Association for the Advancement of Science).

Danish, K., K. Fidler, K. Gallagher, M. Ceronsky and T. Carbonell (2011a) Weekly Climate Change Policy Update: States and Cities—January 31 (Washington, DC: VanNess Feldman) at http://www.vnf.com/news-policyupdates-553.html (cited: 03/03/11).

Danish, K., K. Fidler, K. Gallagher, M. Ceronsky and T. Carbonell (2011b) Weekly Climate Change Policy Update: States June 6–10, 2011 (Washington, DC: VanNess Feldman) at http://www.vnf.com/news-policyupdates-601.html (cited: 06/15/11).

Danish, K., K. Fidler, K. Gallagher, M. Ceronsky and T. Carbonell (2010) United States: Weekly Climate Change Policy Update (New York: Mondaq) at http://www.mondaq.com/unitedstates/article.asp?articleid=99216 (cited 11/29/10).

Danish, K., K. Fidler, K. Gallagher, M. Ceronsky and T. Carbonell (2009) Weekly Climate Change Policy Update: States and Cities—May 26 (Washington, DC: VanNess Feldman) at http://www.vnf.com/news-alerts-364.html (cited: 03/03/11).

De la Vega, A. (2006) Balance y perspectivas del TLCAN en el plano energético. In I. Rousseau (ed.) ¿Hacia la integración de los mercados petroleros en México? (Mexico City: El Colegio de México).

De Place, E. (2008) Inside WCI: Federal Pre-emption. What Happens with a New President? (San Francisco, CA: Architecture for Humanity) at http://www.worldchanging.com/archives/008394.html (cited: 04/07/11).

Dean, M. (1999) Governmentality: Power and Rule in Modern Society (London: SAGE Publications).

Demerse, C. (2008) Canada Blocks Climate Change Progress (Poznan, Poland: Pembina Institute) at http://climate.pembina.org/op-ed/1757 (cited: 04/13/09).

Dente, B. (2010) The Transformations of Regional and Local Governments: Implications for Environmental Policy Integration. In A. Goria, A. Sgobbi and I. von Homeyer (eds.) Governance for the Environment: A Comparative Analysis of Environmental Policy Integration (Cheltenham, UK: Edward Elgar Publishing).

Department of Commerce—US (2013) Regional Data (Washington, DC: Bureau of Economic Analysis) at http://www.bea.gov/regional/gsp/action.cfm (cited: 05/19/13).

Department of Energy—US (2007) North American Energy Working Group (NAEWG) (Washington, DC: U.S. Department of Energy) at http://www.pi.energy.gov/naewg.htm (cited: 03/03/11).

Department of Environment of New York (2011) Adoption of California Vehicle Emissions Standards (New York: Department of Environment) at http://www.dec.ny.gov/energy/43384.html (cited: 03/06/11).

Dingwerth, K. and Ph. Pattberg (2006) Global Governance as a Perspective on World Politics, Global Governance, Vol. 12, No. 2 (April–June), pp. 185–203 (Boulder, CO: Lynne Rienner Publishers).

Djelic, M. and K. Sahlin-Anderson (2006) Transnational Governance: Institutional Dynamics of Regulation (Cambridge, UK: Cambridge University Press).

Domínguez, C. (2011) Beyond Efficiency: The Politics of Investment Policies in the Oil Industry. In Baker Institute Energy Forum, The Future of Oil in Mexico (Mexico City: James A. Baker III Institute for Public Policy and Oxford University).

Domínguez, R. (2009) North American Regionalization: Pragmatic Cooperation as an Alternative to the European Model. In I. Hussain (ed.) North America at the Crossroads: NAFTA after 15 Years (Mexico City: Universidad Iberoamericana, A.C).

Doornbos, M. (2004) "Good Governance": The Pliability of a Political Concept, TRAMES, A Journal of the Humanities and Social Sciences, Vol. 8, No. 4, pp. 372–387 (Tallinn, Estonia: Teaduste Akadeemia Kirjastus).

Dorfman, R. and N. Dorfman (eds.) (1993) Economics of the Environment, Selected Readings, Third Edition (New York: W.W. Norton and Co.).

Douglas, I. (1999) Globalization as Governance: Toward Archaeology of Contemporary Political Reason. In J. Hart and A. Pakrash (eds.) Globalization and Governance (London: Routledge).

Draude, A. (2007) How to Capture Non-Western Forms of Governance, in Favour of an Equivalence Functionalist Observation of Governance in Areas of Limited Statehood (Berlin: Deutsche Forschungsgemeinschaft -SFB 700) at www.sfb-governance.de (cited: 04/20/11).

Drezner, D.W. (2001) Globalization and Policy Convergence (Malden, MA: Blackwell Publishers) at http://www.danieldrezner.com/research/policyconvergence.pdf (cited: 04/20/11).

Duchacek, I. (1988) Perforated Sovereignties and International Relations: Trans-sovereign Contacts of Subnational Governments (Santa Barbara, CA: Greenwood Press).

Duff, D.G. (2008) Carbon Taxation in British Columbia. In The Reality of Carbon Taxes in the 21st Century, Vermont Journal of International Law, Vol. 10, pp. 87–107 (South Royalton, VT: Western Newspaper Publishing).

Dukert, J.M. (2009) Interdependencia energética en América del Norte: el Nuevo Mundo se enfrenta a un nuevo mundo, Revista Mexicana de Política Exterior 87 (Octubre) (México City: Instituto Matías Romero, Secretaría de Relaciones Exteriores).

Dukert, J. (2007) North America. In S. Weintraub, A. Hester and V.R. Prado (eds.), Energy Cooperation in the Western Hemisphere: Benefits and Impediments (Washington, DC: Center for Strategic and International Studies).

Dunn, C. (ed.) (2006) Provinces: Canadian Provincial Politics (Peterborough, ON: Broadview Press).

Durant, R., D. Fiorino and R. O'Leary (eds.) (2004) Environmental Governance Reconsidered: Challenges, Choices and Opportunities (Cambridge: MIT Press).

Elazar, D. (1987) Exploring Federalism (Tuscaloosa, AL: University of Alabama Press).

Ellis, J. and D. Tirpak (2006) Linking GHG Emission Trading Schemes and Markets (Paris: OECD) at http://www.oecd.org/dataoecd/45/35/37672298.pdf (cited: 09/08/10).

Elsenhans, H. (2001) The Political Economy of Good Governance. In Operationalizing Governance for Sustainable Development (Leiden: Koninklijke Brill NV).

Enderlein, H., S. Wälti and M. Zürn (2010) Handbook on Multi-level Governance (Cheltenham, UK: Edward Elgar Publishing).

Endersbee, L. (2008) Global Climate Change has Natural Causes, International Climate and Environmental Change Assessment Project (Washington, DC: ICECAP) at http://icecap.us/images/uploads/Globalclimatechangehasnaturalcauses.pdf (cited: 08/09/10).

Energy Information Administration—US (2013a) Electric Power Monthly (Washington, DC: U.S. Energy Information Administration) at http://www.eia.gov/electricity/monthly/ (cited: 07/02/13).

Energy Information Administration—US (2013b) Electricity Tends to Flow South in North America (Washington, DC: US Energy Information Administration) at http://www.eia.gov/todayinenergy/detail.cfm?id=4270 (cited: 06/20/13).

Energy Information Administration—US (2013c) State Total Energy Rankings (Washington, DC: U.S. Energy Information Administration) at http://www.eia.gov/state/ (cited: 07/15/13).

Energy Information Administration—US (2012a) U.S. Crude Oil, Natural Gas, and NG Liquids Proved Reserves (Washington, DC: U.S. Energy Information) at http://www.eia.gov/naturalgas/crudeoilreserves/ (cited: 03/08/13).

Energy Information Administration—US (2012b) State Electricity Profiles 2010 (Washington, DC: U.S. Energy Information Administration) at http://www.eia.gov/electricity/state/ (cited: 05/10/13).

Energy Information Administration—US (2011) Electric Power Industry—U.S. Electricity Imports from and Electricity Exports to Canada and Mexico 1998–2009 (Washington, DC: U.S. Energy Information Administration) at www.eia.gov/cneaf/electricity/epa/epat6p3.html (cited: 06/09/11).

Energy Information Administration—US (2010a) Annual Energy Outlook 2010 (Washington, DC: U.S. Energy Information Administration) at http://www.eia.gov/oiaf/aeo/pdf/0383(2010).pdf (cited: 06/07/11).

Energy Information Administration—US (2010b) Annual Energy Outlook 2010 with Projections to 2035: American Recovery and Reinvestment Act of 2009: Summary of Provisions (Washington, DC: U.S. Energy Information Administration) at http://www.eia.doe.gov/oiaf/aeo/otheranalysis/aeo_2010analysispapers/arra.html (cited: 01/23/11).

Energy Information Administration—US (2010c) Average Retail Price of Electricity to Ultimate Customers by End-use Sector, by State (Washington, DC: U.S. Energy Information Administration) at http://www.eia.doe.gov (cited 09/01/10).

Energy Information Administration—US (2010d) Short-term Energy Outlook (Washington, DC: U.S. Energy Information Administration) at http://www.eia.gov/emeu/steo/pub/contents.html#CarbonDioxideEmissions (cited: 06/07/11).

Energy Information Administration—US (2008a) Consolidated North American Electric Reliability Corporation Regional Entities (Washington, DC: U.S. Energy Information Administration) at http://www.eia.doe.gov/cneaf/electricity/epa/fig4p2.html (cited: 03/03/11).

Energy Information Administration—US (2008b) Natural Gas Import/Export Pipelines (Washington, DC: Energy Information Administration) at http://www.eia.gov/pub/oil_gas/natural_gas/analysis_publications/ngpipeline/impex.html (cited: 06/10/13).

Enion, R. (2013) Interview on Carbon Markets in North America, April 1 (Los Angeles, CA: University of California in Los Angeles).

Environment Canada (2012) National Inventory Report for Canada 1990–2011 (Gatineau, QC: Environment Canada))

Environment Canada (2008a) National Inventory Report for Canada 1990–2006 (Gatineau, QC: Environment Canada) at http://www.ec.gc.ca/pdb/ghg/inventory_report/2006/som-sum_eng.cfm (cited: 12/03/10).

Environment Canada (2008b) The Path to Progress: A Climate Change Chronology (Gatineau, QC: Environment Canada) at http://www.ec.gc.ca/climate/timeline-e.html (cited: 03/25/10).

Environment News Service (2008) Governor Schwarzenegger Convenes Governors' Global Climate Summit (Lincoln City, OR: Environment News Service) at http://www.ens-ewswire.com/ens/nov2008/2008-11-18-02.html (cited: 07/04/13).

Environment Northeast (2010) Federal Cap and Trade (Ottawa, ON: Environment Northeast) at http://www.env-ne.org/projects/open/p/id/693/program/Climate%20Change%20Solutions (cited: 03/17/11).

Environmental Leader (2010) RGGI Auction Price Falls, Again (Fort Collins, CO: Environmental Leader, Environmental and Energy Management News) at http://www.environmentalleader.com/2010/06/11/rggi-auction-price-falls-again/ (cited: 03/19/11).

Environmental Protection Agency—US (2013) State and Local Climate and Energy Program—California (Washington, DC: U.S. Environmental Protection Agency) at www.epa.gov/statelocalclimate/state/tracking/individual/ca.html (cited: 10/30/13)

Environmental Protection Agency—US (2012) Inventory of U.S. Greenhouse Gas Emissions and Sinks: 1990–2011 (Washington, DC: U.S. Environmental Protection Agency) at http://www.epa.gov/climatechange/Downloads/ghgemissions/US-GHG-Inventory-2013-ES.pdf (cited: 03/26/13).

Environmental Protection Agency—US (2011) Past Climate Change (Washington, DC: U.S. Environmental Protection Agency) at http://www.epa.gov/climatechange/science/pastcc.html (cited: 09/21/11).

Environmental Protection Agency—US (2009) Plain English Guide to the Part 75 Rule (Washington, DC: U.S. Environmental Protection Agency) at http://www.epa.gov/airmarkt/emissions/docs/plain_english_guide_par75_final_rule.pdf (cited: 01/19/11).

Environmental Protection Agency—US (2008) Inventory of U.S. Greenhouse Gas Emissions and Sinks: 1990–2006 (Washington, DC: U.S. Environmental Protection Agency) at http://www.epa.gov/climatechange/emissions/downloads/08_CR.pdf (cited: 07/19/11).

Ethree.com (2007a) CPUC Greenhouse Gas Modeling Renewable Portfolio Renewable Portfolio Standards—Assumptions (San Francisco, CA and Vancouver, BC: Ethree.com) at http//www.ethree.com/./7%202020%20Base%20Case%20Input%20Summary%20v4.doc (cited: 11/20/10).

Ethree.com (2007b) The Role of Renewable Portfolio Standards in the GHG Model (San Francisco, CA and Vancouver, BC: Ethree.com) at http://www.ethree.com/GHG/10%20RPS%20Assumptions%20v2.doc (cited: 11/20/10).

European Climate Exchange (2009) Database (Copenhagen: European Environmental Agency) at www.europeanclimateexchange.com (cited: 01/28/09).

Evans, P.B. (1997) The Eclipse of the State? Reflections on Stateness in an Era of Globalization, World Politics, A Quarterly Journal of International Relations, Vol. 50, No. 1, pp. 62–87 (Washington, DC: Johns Hopkins University Press).

Evans, P.B., D. Rieschemeyer and Th. Scokpol (1985) Bringing the State Back In (Cambridge, UK: Cambridge University Press).

Falkner, Robert (ed.) (2013) The Handbook of Global Climate and Environmental Policy (Chichester, UK: John Wiley & Sons, Ltd).

Farnsworth, D. (2010) Overview of the Regional Greenhouse Gas Initiative (RGGI) (Montpelier, VT: Vermont Public Service Board).

Farrell, H. and A. Héritier (2003) Formal and Informal Institutions under Codecision: Continuous Constitution-building in Europe, Governance: A International Journal of Policy, Administration and Institutions, Vol. 16, No. 4 (October), pp. 577–600 (Oxford: Blackwell Publishing) at http://www.utsc.utoronto.ca/~farrell/governance.pdf (cited: 06/15/11).

Federal Energy Regulatory Commission (2010) Market Oversight (Washington, DC: FERC) at http://www.ferc.gov/market-oversight/othr-mkts/emiss-allow/2010/03-2010-othr-emns-no-so-pr.pdf (07/10/11).

Ferejohn, J. and B.R. Weingast (1997) The New Federalism: Can the States Be Trusted? (Stanford, CA: Hoover Institution Press).

Ferreira, H. (2005) Construir las regiones (Mexico City: Centro Lindavista, A.C.).

Fiorino, Daniel J. 2004. Flexibility. In R. Durant, D. Fiorino and R. O'Leary (eds.) Environmental Governance Reconsidered: Challenges, Choices and Opportunities (Cambridge, MA: MIT Press).

Flachsland, Ch., O. Edenhofer, M. Jakob and J. Steckel (2008) Developing the International Carbon Market: Linking Options for the EU ETS (Potsdam: Potsdam Institute for Climate Impact Research) at http://www.joanneum.at/climate/linking/PIK_Linking_new.pdf (cited: 06/17/11).

Foucault, M. (1975) Garzón, A. (trans.) (1998) Vigilar y castigar: Nacimiento de una prisión (México: Editorial Siglo XXI).

Freestone, D. and Ch. Streck (2009) Legal Aspects of Carbon Trading: Kyoto, Copenhagen, and Beyond (Oxford, UK: Oxford University Press).

Friedman, L. (2010) Cancun Agreements Preserves an Escape Hatch for Japan and Other Industrial Nations, The New York Times (December 17) (New York: New York Times) at http://www.nytimes.com/cwire/2010/12/17/17climatewirecancun-agreement-preserves-an-escape-hatch-f-38242.html?pagewanted=all (cited: 07/02/11).

Fry, E.H. (2004) The Role of Subnational Governments in the Governance of North America, Mapping the New North American Reality, Working Paper Series, no. 2004-09d (Montreal, QC: Institute for Research on Public Policy).

Gallup USA (2013) Americans' Concerns about Global Warming on the Rise (Princeton, NJ: Gallup USA) at http://www.gallup.com/poll/161645/americans-concerns-global-warming-rise.aspx?ref=more (cited: 06/17/13).

Gardner, T. and P. Henderson (2010) FACTBOX: Regional US Carbon Markets Survive Elections (San Francisco, CA: Thompson-Reuters) at http://www.reuters.com/article/2010/11/03/us-usa-carbon-markets-idUSTRE6A261N20101103 (cited: 02/02/11).

Gardner, T. and P. Henderson (2009) Green States Line Up Behind U.S. Climate Bill (San Francisco, CA: Thompson-Reuters) at http://www.reuters.com/article/environmentNews/ idUSTRE55P6KR20090627 (cited: 02/02/11).

Georgetown Climate Center (2013) Challenge to New York's Participation in RGGI Dismissed (Washington, DC: Georgetown Climate Center) at http://www.georgetownclimate.com/challenge-to-new-york-s-participation-in-rggi-dismissed (cited: 07/03/13).

Gerdes, J. (2013) How will California Slash Greenhouse Gas Emissions 80% by 2050? (New York: Forbes) at http://www.forbes.com/sites/justingerdes/2013/02/27/how-will-california-slash-greenhouse-gas-emissions-80-by-2050/ (cited: 07/02/13).

Getis, A., J. Getis and J. Fellman (1991) Introduction to Geography (Dubuque, IA: Wm. C. Brown Publishers).

Gibbons, J.H. and P.D. Blair (1991) US Energy Transition: On Getting from Here to There, Physics Today (July) (New York: American Institute of Physics Publishing LLC).

Giddens, A. (2010) La Política del Cambio Climático (Madrid: Alianza Editorial).

Goldthau, A. (2013) The Handbook of Global Energy Policy (Malden, MA: Wiley-Blackwell).

Gopalakrishnan, Ch. (ed.) (2000) Classic Papers in Natural Resource Economics (New York: Palgrave Macmillan).

Goria, A., A. Sgobbi and I. Homeyer (2010) Governance for the Environment: A Comparative Analysis of Environmental Policy Integration (Cheltenham, UK: Edward Elgar Publishing).

Government of British Columbia (2008) Climate Action Plan (Vancouver, BC: Government of British Columbia) at http://www.livesmartbc.ca/attachments/climateaction_plan_web.pdf (cited: 09/23/10).

Government of British Columbia (2002) BC Energy Plan: A Vision for Clean Energy Leadership (Vancouver, BC: Government of British Columbia) at http://www.energyplan.gov.bc.ca/bcep/default.aspx?hash=6 (cited: 12/08/10).

Government of Canada (2012) Canada–US Energy Relations (Ottawa, ON: Government of Canada) at http://www.canadainternational.gc.ca/san_diego/bilateral_relations_bilaterales/energy-energie.aspx?lang=eng&view=d (cited: 03/15/13).

Government of Manitoba (2008a) C.C.S.M. c. C135—The Climate Change and Emissions Reductions Act (Winnipeg, MB: Government of Manitoba) at http://web2.gov.mb.ca/laws/statutes/ccsm/c135e.php (cited 12/16/10).

Government of Manitoba (2008b) Executive Summary: Meeting Our Commitments – Reducing Emissions in Manitoba and Beyond (Winnipeg, MB: Government of Manitoba) at http://www.gov.mb.ca/asset_library/en/beyond_kyoto/executive_summary.pdf (cited 12/03/10).

Government of Massachusetts (2010) Massachusetts Clean Energy and Climate Plan for 2020 (Boston, MA: Government of Massachusetts) at http://www.mass.gov/Eoeea/docs/eea/energy/2020-clean-energy-plan.pCity (consulted: 01/12/11).

Government of New Brunswick (2009) New Brunswick and Maine Support International Energy Corridor, NB 380 (Saint John, NB: Government of New Brunswick) at http://www.gnb.ca/cnb/news/ene/2009e0380en.htm (cited: 01/12/11).

Government of Ontario (2008) Ontario Climate Action Plan (Ottawa, ON: Government of Ontario) at http://www.ene.gov.on.ca/en/air/climatechange/ourplan.php (cited 08/26/10).

Government of Ontario (2007) Ontario's Drive Clean: A Mandatory, Vehicle Emissions-testing Program (Ottawa, ON: Government of Ontario) at http://www.ene.gov.on.ca/cons/3717e07.pdf (cited 12/08/10).

Government of Quebec (2013a) Regulation Respecting a Cap-and-trade System for Greenhouse Gas Emission Allowances: A Technical Overview (Montreal, QC: Government of Quebec) at http://www.mddefp.gouv.qc.ca/changements/carbone/SPEDE-description-technique-en.pdf (cited: 07/02/13).

Government of Quebec (2013b) Regulation Respecting a Cap-and-trade System for Greenhouse Gas Emission Allowances: Updated Version of Documents of 2011 and 2012 (Montreal, QC: Government of Quebec) at http://www2.publicationsduquebec.gouv.qc.ca/dynamicSearch/telecharge.php?type=3&file=/Q_2/Q2R46_1_A.HTM (cited: 07/02/13).

Government of Quebec (2012) Regulation Respecting the Delegation of Management of Certain Parts of a Cap-and-trade System for Greenhouse Gas Emission Allowances (Montreal, QC: Government of Quebec) at http://www2.publicationsduquebec.gouv.qc.ca/dynamicSearch/telecharge.php?type=1&file=2012A%2F2390.PDF (cited: 07/02/13).

Government of Quebec (2008) Québec and Climate Change: 2006–2012 Action Plan (Quebec City, QC: Government of Quebec) at http://www.mddep.gouv.qc.ca/changements/plan_action/2006-2012_en.pdf (cited: 05/17/10).

Government of Quebec (2007) Using Energy to Build the Québec of Tomorrow: Québec Energy Strategy 2006–2015—Summary (Quebec City, QC: Government of Quebec) at http://www.mrnf.gouv.qc.ca/english/publications/energy/strategy/energy-strategy-2006-2015-summary.pdf (cited 04/23/10).

Governor's Global Climate Summit (2009) Greening the Summit (California: Green Technology Smart Media) at http://climatesummit.greenstreetscene.com/channel/conferences/globalclimatesummit2/summit-overview/idx0-1-governors-global-climate-summit-2?event_id=319 (08/11/12).

Grady, P. and A. Sharpe (eds.) (2001) The State of Economics in Canada: Festschrift in Honor of David Slater (Ottawa, ON: Center for the Study of Living Standards).

Grimes, K. (2012) CA Energy Schemes: "We are getting fleeced" (Sacramento, CA: CalWatchDog) at http://www.calwatchdog.com/2012/04/05/ca-energy-schemes-we-are-getting-fleeced/ (cited: 01/07/13).

Guertin, D.L., W.K. Davis and J.E. Gray (1992) U.S. Energy Imperatives for the 1990s: Leadership, Efficiency, Environmental Responsibility, and Sustained Economic Growth (Lanham, MD: University Press of America, Inc.).

Guha-Knasnobis, B., R. Kanbur and E. Ostrom (2006) Linking the Formal and Informal Economy: Concepts and Policies (Oxford: Oxford University Press, UNU-WIDER Studies in Development Economics, and EGDI) at http://www.oecd.org/dataoecd/52/25/37791314.pdf (cited: 07/07/11).

Gunderson, M. (2001) North American Economic Integration and Globalization. In P. Grady and A. Sharpe (eds.) The State of Economics in Canada: Festschrift in Honor of David Slater (Ottawa, ON: Center for the Study of Living Standards).

Haas, P. (1992) Introduction: Epistemic Communities and International Policy Coordination, International Organization, Vol. 46, No. 1 (Cambridge, UK: Cambridge University Press) at http://fiesta.bren.ucsb.edu/~gsd/595e/docs/13.%20Haas_Intro_Epistemic%20Communities_plus.pdf (cited: 07/23/11).

Haigh, J. (2011) Solar Influences on Climate, Briefing Paper No 5. Executive Summary (London: Grantham Institute for Climate Change Imperial College).

Hall, P. (2003) Ontology and Methodology in Comparative Research. In J. Mahoney and D. Ruechemeyer (eds.) Comparative Historical Analysis in the Social Sciences (Cambridge, UK: Cambridge University Press).

Hall, P. and K. Thelen (2009) Institutional Change in Varieties of Capitalism, Socio-economic Review, Vol. 7, No. 1, pp. 7–34 (Oxford: Oxford University Press).

Hall, R.B. and Th. Biersteker (eds.) (2002) The Emergence of Private Authority in Global Governance (Cambridge, UK: Cambridge University Press).

Hanley, N., J. Shogren and B. White (1997) Environmental Economics in Theory and Practice (Oxford: Oxford University Press).

Hardin, G. (1993) The Tragedy of the Commons. In R. Dorfman and N. Dorfman (eds.) (1993) Economics of the Environment, Selected Readings, Third Edition (New York: W.W. Norton and Co.).

Hardt, M. and A. Negri (2000) Empire (Cambridge, MA: Harvard University Press).
Harrison, K. (2010a) The Struggle of Ideas and Self-interest in Canadian Climate Policy. In K. Harrison and L. McIntosh (eds.) Global Commons, Domestic Decisions: The Comparative Politics of Climate Change. Cambridge, MA: MIT Press.
Harrison, K. (2010b) The United States as Outlier: Economic and Institutional Challenges to US Climate Policy. In K. Harrison and L. McIntosh (eds.) Global Commons, Domestic Decisions: The Comparative Politics of Climate Change. Cambridge, MA: MIT Press.
Harrison, K. (2007) The Road Not Taken: Climate Change Policy in Canada and the United States, Global Environmental Politics, Vol. 7, No. 4 (November), pp. 92–117 (Cambridge, MA: Massachusetts Institute of Technology).
Harrison, K. (2006) The Struggle between Ideas and Self-interest: Canada's Ratification and Implementation of the Kyoto Protocol, Working Paper, International Studies Association (San Diego, CA: International Studies Association).
Harrison, K. and L. McIntosh (eds.) (2010) Global Commons, Domestic Decisions: The Comparative Politics of Climate Change (Cambridge, MA: MIT Press).
Hart, J. and A. Pakrash (eds.) (1999) Globalization and Governance (London: Routledge).
Hasenclever, A., P. Mayer and V. Rittberger (1997) Theories of International Regimes (Cambridge, UK: Cambridge University Press).
Hedges, A. (2009) The Secondary Market for Emissions Trading: Balancing Market Design and Market Based Transaction Norms. In D. Freestone and Ch. Streck (eds.) Legal Aspects of Carbon Trading: Kyoto, Copenhagen, and Beyond (Oxford, UK: Oxford University Press).
Held, D. and A. McGrew (eds.) (2002) Governing Globalization: Power, Authority and Global Governance (Cambridge, UK: Polity Press).
Held, D. and A. McGrew (eds.) (2000) The Global Transformations Reader: An Introduction to the Globalization Debate (Cambridge, UK: Polity Press).
Held, D., A. McGrew, D. Goldblatt and J. Perraton (1999) Global Transformations: Politics, Economics and Culture (Cambridge, UK: Polity Press).
Helmke, G. and S. Levitsky (2004) Informal Institutions and Comparative Politics: A Research Agenda, International Seminar: Informal Institutions and Development (Paris: OECD) at http://www.oecd.org/dataoecd/52/38/37790183.pdf (cited: 02/07/11).
Hettne, B. (2004) The New Regionalism Revisited. In F. Söderbaum and T. M. Shaw (eds.) Theories of New Regionalism (Basingstoke: Palgrave Macmillan).
Hettne, B. (2000) Global Market versus Regionalism. In D. Held and A. McGrew (eds.) The Global Transformations Reader: An Introduction to the Globalization Debate (Cambridge, UK: Polity Press).
Hiernaux, D. (1993) El concepto de espacio y el análisis regional, Revista Secuencias, Vol. 25, pp. 89–110 (Mexico City: Instituto José María Luis Mora).

Hiskes, J. (2010) The Post-election Outlook for Regional Cap and Trade, Grist Organization—The Climate Desk (Seattle, WA: Grist—The Climate Desk) at http://motherjones.com/blue-marble/2010/11/post-election-outlook-regional-cap-and-trade (cited: 02/12/11).

Hocking, B. (1999) Patrolling the “Frontier”: Globalization, Localization and the “Actorness” of Non-central Governments. In F. Aldecoa and M. Keating (eds.) Paradiplomacy in Action: The Foreign Relations of Subnational Governments (London: Frank Cass Publishers).

Hoffert, M.I., K. Caldeira, G. Benford, D. Criswell, C. Green, H. Herzog, A. Jain, H. Kheshgi, K. Lackner, J. Lewis, H. Lightfoot, W. Manheimer, J. Mankins, M. Mauel, L. Perkins, M. Schlesinger, T. Volk and T. Wigley (2002) Advanced Technology Paths to Global Climate Stability: Energy for a Greenhouse Planet, Science, Vol. 298 (Washington, DC: American Association for the Advancement of Science).

Hofman, S.C. and F. Mérand (2012) Regional Organizations à la carte: The Effects of Institutional Elasticity. In T. V. Paul (ed.) International Relations Theory and Regional Transformation (Cambridge, UK: Cambridge University Press).

Holzinger, K. and C. Knill (2008) Theoretical Framework: Causal Factors and Convergence Expectations. In K.C.K. Holzinger and B. Arts (eds.) Environmental Policy Convergence in Europe: The Impact of International Institutions and Trade (Cambridge, UK: Cambridge University Press).

Holzinger, K., H. Jörgens and C. Knill (2008) State of the Art: Conceptualising Environmental Policy Convergence. In K.C.K. Holzinger and B. Arts (eds.) Environmental Policy Convergence in Europe: The Impact of International Institutions and Trade (Cambridge, UK: Cambridge University Press).

Holzinger, K.C.K. and B. Arts (eds.) (2008) Environmental Policy Convergence in Europe: The Impact of International Institutions and Trade (Cambridge, UK: Cambridge University Press).

Hooghe, L. and G. Marks (2003) Unraveling the Central State, but How? Types of Multi-level Governance, The American Political Science Review, Vol. 97, No. 2 (May), pp. 233–243 (Princeton, NJ: American Political Science Association) at http://www.jstor.org/stable/3118206 (cited: 12/14/09).

Hoppkins, M. (1999) The Planetary Bargain: Corporate Social Responsibility Comes of Age (New York: St. Martin’s Press).

Hristoulas, A., C. Denis and D. Wood (eds.) (2005) Canadá: política y gobierno en el siglo XXI (Mexico City: ITAM-Porrúa-Cámara de Diputados LIX Legislatura).

Hussain, I. (ed.) (2009) North America at the Crossroads: NAFTA after 15 Years (Mexico City: Universidad Iberoamericana, A.C.).

Hutchinson, G.E. (1965) The Ecological Theater and the Evolutionary Play (New Haven, CT: Yale University Press).

Hydro-Québec (2011) Comparison of Electricity Prices in Major North American Cities (Quebec City: Hydro-Québec) at http://www.hydroquebec.com/publications/en/comparison_prices/ (cited: 07/01/13).

Hydro-Québec (2010) Découvrir Hydro-Québec TransÉnergie et son réseau. Notre réseau en bref (Quebec City: Hydro-Québec) at http://www.hydroquebec.com/transenergie/fr/reseau/bref.html (cited: 06/07/11).

Ibarrarán, M. and R. Boyd (2007) Hacia el futuro: Energy, Economics, and the Environment in 21st century Mexico (Dordrecht: Springer).

Instituto Nacional de Ecología (2012) Quinta Comunicación Nacional ante la Convención Marco de las Naciones Unidas sobre el Cambio Climático (Mexico City: Instituto Nacional de Ecología) at http://www2.inecc.gob.mx/publicaciones/consultaPublicacion.html?id_pub=685 (cited: 06/02/13).

Instituto Nacional de Ecología (2011) El cambio climático y los gases de efecto invernadero (Mexico City: Instituto Nacional de Ecología) at http://cambio_climatico.ine.gob.mx/pregfrecuentes.html#2 (cited: 05/19/11).

Instituto Nacional de Ecología (2010) Resumen Ejecutivo. México Cuarta Comunicación Nacional Mexico City: Instituto Nacional de Ecología) at http://www2.ine.gob.mx/publicaciones/libros/615/resumen.pdf (cited 12/11/10).

Instituto Nacional de Ecología (2009) Cuarta Comunicación Nacional ante la Convención Marco de las Naciones Unidas sobre el Cambio Climático (Mexico City: Instituto Nacional de Ecología) at http://www2.ine.gob.mx/publicaciones/consultaPublicacion.html?id_pub=615 (cited: 04/11/12).

Instituto Nacional de Ecología (2008) Tercera Comunicación Nacional sobre Cambio Climático (1990–2002) (Mexico City: Instituto Nacional de Ecología) at http://www.ine.gob.mx/cclimatico/comnal3.html (cited: 11/25/08).

Intergovernmental Panel on Climate Change (2010) Database (Geneva: World Meteorological Organization/IPCC) at www.ipcc.ch (cited: 04/29/11).

Intergovernmental Panel on Climate Change (2007a) Working Group I Report: The Physical Science Basis (Geneva: World Meteorological Organization/International Panel on Climate Change) at http://www.ipcc.ch/publications_and_data/publications_ipcc_fourth_assessment_report_wg1_report_the_physical_science_basis.htm (cited: 03/18/10).

Intergovernmental Panel on Climate Change (2007b) Working Group II Report: Impacts, Adaptation and Vulnerability (Geneva: World Meteorological Organization/International Panel on Climate Change) at http://www.ipcc.ch/publications_and_data/publications_ipcc_fourth_assessment_report_wg2_report_impacts_adaptation_and_vulnerability.htm (cited: 03/18/10).

Intergovernmental Panel on Climate Change (2007c) Working Group III Report: Mitigation of Climate Change (Geneva: World Meteorological Organization/International Panel on Climate Change) at http://www.ipcc.ch/publications_and_data/publications_ipcc_fourth_assessment_report_wg3_report_mitigation_of_climate_change.htm (cited: 03/18/10).

International Energy Agency (2011) Prospect of Limiting the Global Increase in Temperature to 2°C is Getting Bleaker (Paris: OECD) at http://www.iea.org/index_info.asp?id=1959 (cited: 07/15/11).

International Energy Agency (2010) Global Energy Outlook—Summary (Paris: OECD) at http://www.iea.org/Textbase/npsum/weo2010sum.pCity (cited: 01/24/11).

International Council for Local Environmental Initiatives (2013a) ICLEI realiza evento paralelo para gobiernos locales en COP18 (Mexico City: ICLEI Mexico) at http://www.iclei.org.mx/web/index.php/noticias/ICLEI_panel_desarrollo_sustentable_COP18 (cited: 06/13/13).

International Council for Local Environmental Initiatives (2013b) Pioneering Cities and Experts Meet to Further Implementation of Durban Adaptation Charter (Cape Town: ICLEI Africa) at http://www.iclei.org/details/browse/2/article/pioneering-cities-and-experts-meet-to-further-implementation-of-durban-adaptation-charter.html?tx_ttnews%5BbackPid%5D=22&cHash=a9995f42597300f380cb1cdd2a5552e9 (cited: 06/13/13).

International Council for Local Environmental Initiatives (2011a) Durban Adaptation Charter for Local Governments (Cape Town: ICLEI Africa) at http://www.durbanadaptationcharter.org/ (cited: 06/13/13).

International Council for Local Environmental Initiatives (2011b) HEAT+ Harmonized Emissions Analysis Tool (Bonn: ICLEI) at http://heat.iclei.org/heatplusv4n/downloads/HEATplus_brochure.pdf (cited: 06/13/13).

International Council for Local Environmental Initiatives (2010a) Día 10: Miércoles 8 de Diciembre de 2010: Se reconoce oficialmente la importancia de los gobiernos locales como piezas clave en la lucha contra el cambio climático dentro de las sesiones de la ONU en la COP-16 (Cancun, MX: ICLEI) at http://www.iclei.org.mx/web/index.php/seccion/COP-16 (cited: 01/05/11).

International Council for Local Environmental Initiatives (2010b) Local Government Climate Roadmap. Daily Report-Day 2 (Cancun, MX: ICLEI) at http://iclei.org/fileadmin/template/project_templates/climate-roadmap/files/COP16/Daily_Report/ daily_ report_30nov2010.pdf (cited: 01/05/11).

International Council for Local Environmental Initiatives (2010c) Local Government Road Map (Cancun, MX: ICLEI) at http://www.iclei.org/index.php?id=11817 (cited: 01/05/11).

International Council for Local Environmental Initiatives (2010d) The ICLEI Story (Bonn: ICLEI) at http://www.iclei.org/index.php?id=748 (cited: 05/13/10).

International Council for Local Environmental Initiatives (2009) 1st European Local Government Positioning Summary on Climate Change and Energy (Bonn: ICLEI) at http://www.iclei-europe.org/fileadmin/template/projects/lg-action/files/en/1st_LG_Positioning_summary_Summit_June_09.pdf (cited: 05/17/10).

International Energy Agency (2012a) CO2 Emissions from Fuel Combustion Highlights (Paris: International Energy Agency) at http://www.iea.org/publications/freepublications/publication/CO2Highlights2012.XLS (cited: 08/07/13).

International Energy Agency (2012b) World Energy Outlook 2012: Executive Summary (Paris: International Energy Agency) at http://www.iea.org/Textbase/npsum/weo2012sum.pdf (cited: 02/18/13).

Jaccard, M., N. Rivers and J. Simpson (2007) Hot Air: Meeting Canada's Climate Change Challenge (Toronto, ON: McClelland & Stewart).

Jacob, D. (2005) Radiative Forcing of Climate Change: Expanding the Concept and Addressing Uncertainties (Washington, DC: Committee on Radiative Forcing Effects on Climate Research Committee Board on Atmospheric Sciences and Climate Division on Earth and Life Studies and The National Academies Press) at http://www.nap.edu/catalog/11175.html (cited: 10/30/13).

Jagers, S. and J. Stripple (2003) Climate Governance Beyond the State, Global Governance, Vol. 9, No. 3, pp. 385–399 (Boulder, CO: Lynne Rienner Publishers).

Jelin, E. (ed.) (2003) Más allá de la nación: las escalas múltiples de los movimientos sociales (Buenos Aires: Libros del Zorzal).

Jevons, W.S. (1865) The Coal Question: An Inquiry Concerning the Progress of the Nation, and the Probable Exhaustion of Our Coal-mines (London: Macmillan and Co.).

Johnson, B. (2010) Governor Races: Climate Deniers Threaten the Northeast RGGI Climate Compact (Washington, DC: Center for American Progress Action Fund) at http://thinkprogress.org/green/2010/10/15/174817/gop-governor-deniers-iii/ (cited: 02/19/11).

Jordan, A., R.K. Wurzel and A. Zito (2005) The Rise of "New" Policy Instruments in Comparative Perspective: Has Governance Eclipsed Government?, Political Studies, Vol. 53, No. 10, pp. 477–496 (Hoboken, NJ: John Wiley & Sons, Inc.).

Juillet, L. (2005) La política ambiental canadiense. In A. Hristoulas, C. Denis and D. Wood (eds.) Canadá: política y gobierno en el siglo XXI (Mexico City: ITAM-Porrúa-Cámara de Diputados LIX Legislatura).

Juillet, L. and M. Patterson (2009) Une difficile modernisation écologique: l'evolution de la politique canadienne sur les changements climatiques, Les politiques publiques au Canada, pouvoir, conflits et idéologie (Quebec: Presses de l'université de Laval).

Kahn, J. (2005) The Economic Approach to Environmental and Natural Resource Economics, Third Edition (Mason, OH: Thomson South-Western).

Kaul, I. and R. Mendoza (2003) Advancing the Concept of Public Goods. In I. Kaul, P. Conceição, K. Le Goulven and R. Mendoza, Providing Global Public Goods: Managing Globalization (New York: Oxford University Press).

Kaul, I., I. Grundberg and M. Stern (1999) Global Public Goods, Concepts, Policies and Strategies. In I. Kaul, I. Grundberg and M. Stern (eds.) Global Public Goods: International Cooperation in the 21st Century (New York: Oxford University Press).

Kaul, I., P. Conceição, K. Le Goulven and R. Mendoza (eds.) (2003) Providing Global Public Goods: Managing Globalization (New York: Oxford University Press).

Keohane, R. (2002) Power and Governance in a Partially Globalized World (London: Routledge).

Keohane, R. (1998) International Institutions: Can Interdependence Work?, Foreign Policy, Vol. 110, Special Edition: Frontiers of Knowledge (Spring), pp. 82–96 (Washington, DC: Washington Post Newsweek Interactive, LLC).

Keohane, R. and J.S. Nye (2000) Introduction. In J.S. Nye and J. Donahue (eds.) Governance in a Globalizing World (Washington, DC: Brookings Institution Press).

Keohane, R. and E. Ostrom (eds.) (1995) Local Commons and Global Interdependence (London: SAGE Publications).

Kern, K. (2010) Governing Climate Change in Cities: Modes of Urban Climate Governance in Multi-level Systems (Paris: OECD, Wageningen University and Gotelind Alber Sustainable Energy and Climate Policy) at http://www.oecd.org/dataoecd/54/63/42545036.pdf (cited: 03/23/11).

Kettl, D. (2002) Environmental Governance: A New Approach of Environmental Policy (Washington, DC: Brookings Institution).

Kibel, P.S. (2001) The Paper Tiger Awakens: North American Environmental Law after the Cozumel Reef Case, Columbia Journal of Transnational Law, Vol. 395, No. 39 (New York: Columbia University Press).

Kincaid, J. (1998) The International Competence of US States and Their Local Governments, Journal of Regional and Federal Studies, Vol. 9, No. 1, pp. 111–130 (Portland: Frank Cass Journals).

Knigge, M. and C. Bausch (2006) Climate Change Policies at the U.S. Subnational Level: Evidence and Implications (London: Institute for International and European Environmental Policy) at www.ecologic.de (cited 01/03/11).

Knight, J. and R. Calvert (1995) The Rational Choice Theory of Social Institutions: Cooperation, Coordination, and Communication. In J.S. Banks and E.A. Hanushek (eds.) Modern Political Economy (New York: Cambridge University Press).

Knopman, D. and E. Fleschner (1999) Second Generation of Environmental Stewardship: Improve Environmental Results and Broaden Citizen Engagement-briefing (Washington, DC: Progressive Policy Institute).

Kobrin, S. (2008) Globalization, Transnational Corporations and the Future of Global Governance. In A.G. Scherer and G. Palazzo (eds.) Handbook of Research on Global Corporate Citizenship (Cheltenham, UK and Northampton, MA: Edward Elgar Publishing).

Kooiman, J. (2000) Societal Governance: Levels, Models, and Orders of Social–Political Interaction. In J. Pierre (coord.) Debating Governance: Authority, Steering and Democracy (Oxford: Oxford University Press).

Krahmann, E. (2003) National, Regional & Global Governance, Global Governance, Vol. 9, No. 3, pp. 323–346 (Boulder, CO: Lynne Rienner Publishers).

Krasner, S. (1983) International Regimes (New York: Cornell University Press).

Lagerkvist, M. (2010) Why RGGI is Failing & Unlikely to Cut CO-2 Emissions Before 2030: Investigative Report (Red Bank, NJ: New Jersey Watchdog) at http://newjersey.watchdog.org/2010/11/17/why-rggi-may-fail-to-impact-co-2-emissions-for-20-years/ (cited: 03/02/11).

LaMotte, K.R., D.M. Williamson and L.A. Hopkins (2009) Emissions Trading in the US: Legal Issues. In D. Freestone and Ch. Streck (eds.) Legal Aspects of Carbon Trading: Kyoto, Copenhagen, and Beyond (Oxford, UK: Oxford University Press).

Larner, W. and W. Walters (2004) Global Governmentality: Governing International Spaces (London: Routledge).

Larsen, J., Th. Damassa and R. Levinson (2007) Charting the Midwest: An Inventory and Analysis of Greenhouse Gas Emissions in America's Heartland (Washington, DC: World Resources Institute) at http://www.wri.org/publication/charting-the-midwest (cited 01/11/11).

Lee, M. (2011) Fair and Effective Carbon Pricing: Lessons from BC (Victoria, BC: Canadian Centre for Policy Alternatives and Sierra Club B.C.) at http://www.policyalternatives.ca/sites/default/files/uploads/publications/BC%20Office/2011/02/CCPA-BC_Fair_Effective_Carbon_FULL_0.pdf (cited: 02/24/11).

Li, Sh., S. Ho Park and S. Li (2003) The Great Leap Forward: The Transition from Relation-based Governance to Rule-based Governance, Organisational Dynamics, Vol. 33, No. 1, pp. 63–78 (London: Elsevier Ltd).

Linowes, L. (2010) Regional Greenhouse Gas Initiative (RGGI): A Cap-Tax-Spend Model to NOT Follow, Master Resource: A Free-market Energy Blog at http://www.masterresource.org/2010/07/rggi-cap-tax-spend-model/ (cited: 07/03/13).

Lipschutz, R.D. and C. Fogel (2002) Regulation for the Rest of Us? Global Civil Society and the Privatization of Transnational Regulation. In Th. J. Biersteker and R.B. Hall (eds.) The Emergence of Private Authority in Global Governance (Cambridge, UK: University of Cambridge Press).

Litfin, K. (2000) Environment, Wealth, and Authority: Global Climate Change and Emerging Modes of Legitimation, International Studies Review, Vol. 2, No. 2, pp. 119–148 (Hoboken, NJ: John Wiley & Sons, Inc.) at http://www3.interscience.wiley.com/journal/119052072/issue (cited: 07/03/11).

Lomolino, M., B. Riddle and J. Brown (2006) Biogeography, Third Edition (Sunderland, MA: Sinauer Associates, Inc.).

Lucas, A.R. (2004) Legal Constraints and Opportunities: Climate Change and the Law. In H. Coward and A. Weaver (eds.) Hard Choices: Climate Change in Canada (Waterloo, ON: Wilfrid Laurier University Press).

Lum, R. and K. Knutson (2013) North American Investment in Electric Transmission to Top $14bn in 2013 (Tulsa, OK: PennWell Corporation) at http://www.elp.com/articles/electric-light-and-power-newsletter/articles/2013/01/north-american-transmission-investment-to-top--14-billion-in-201.html (cited: 06/10/13).

MacKay, W. (2004) Canadian Federalism and the Environment: The Literature, Georgetown International Environmental Law Review (Fall) (Washington, DC: Georgetown University).

Mahoney, J. and D. Ruechemeyer (eds.) (2001) Comparative Historical Analysis in the Social Sciences (Cambridge, UK: Cambridge University Press).

Mann, M. (2000) Has Globalization Ended the Rise and Rise of the Nation-state? In D. Held and A. McGrew (eds.) The Global Transformations Reader (Cambridge, UK: Polity Press).

Mansell, M., P. Sopher and C. Munnings (2013) Québec—The World's Carbon Markets: A Case Study Guide to Emissions Trading (Washington, DC: International Emissions Trading Association and Environmental Defense Fund) at http://www.ieta.org/worldscarbonmarkets (cited: 07/02/13).

Mansfield, E.E. and H.V. Milner (1999) The New Wave of Regionalism, International Organization, Vol. 53, No. 3 (Summer), pp. 589–627 (Cambridge, UK: Cambridge University Press).

Maps of the World (2011) Energy Resources in North America (San Jose, CA: Maps of the World) at http://www.mapsofworld.com/north-america/energy-resources.html (cited: 06/14/11).

Martin, Lisa (1995) Heterogeneity, Linkage and Commons Problems. In R. Keohane and E. Ostrom (1995) (eds.) Local Commons and Global Interdependence (London: SAGE Publications).

Marsh, D. and M. Smith (2000) Understanding Policy Networks: Towards a Dialectical Approach, Political Studies, Vol. 48, No. 1, pp. 4–21 (Hoboken, NJ: John Wiley & Sons, Inc.) at http://www.blackwell-synergy.com/doi/abs/10.1111/1467-9248.00247 (cited: 10/30/13).

Masini, F. (2011) Economic Epistemology and Methodological Nationalism: A Federalist Perspective, Perspectives on Federalism, Vol. 3, No. 1 (Torino: Centro Studi sul Federalismo) at http://www.on-federalism.eu/attachments/090_download.pdf (cited: 18/05/11).

McAdam, D., S. Tarrow and Ch. Tilly (2001) Dynamics of Contention (New York and Cambridge, UK: Cambridge University Press).

McCright, A.M. and R.E. Dulap (2003) Defeating Kyoto: The Conservative Movement's Impact on U.S. Climate Change Policy, Social Problems, Vol. 50, No. 3 (August), pp. 348–373 (Berkeley, CA: Society for the Study of Social Problems) at http://ireswb.cc.ku.edu/~crgc/NSFWorkshop/Readings/Defeating%20Kyoto.pdf (cited: 04/06/11).

McKinsey & Company (2007) Reducing US Greenhouse Gas Emissions: How Much at What Cost? (New York: McKinsey & Company) at http://www.mckinsey.com/en/Client_Service/Sustainability/Latest_thinking/Costcurves.aspx (cited: 10/05/11).

McKinstry, R.B., J.C. Dernbach and Th. D. Peterson (2008) Federal Climate Change Legislation as if the States Matter, Natural Resources & Environment, Vol. 22, No. 3 (Winter) (Chicago, IL: American Bar Association).

Meckling, J. and C. Hepburn (2013) Economic Instruments for Climate Change. In R. Falkner (ed.) The Handbook of Global Climate and Environmental Policy (Chichester, UK: John Wiley & Sons, Ltd).

Mehling, M. (2009) Linking Emissions Trading Schemes. In D. Freestone and C. Streck (eds.) Legal Aspects of Carbon Trading: Kyoto, Copenhagen, and Beyond (Oxford, UK: Oxford University Press).

Méndez, J.L. (ed.) (2010) Los grandes problemas de México: XIII Políticas Públicas, Edición Abreviada (Mexico City: El Colegio de México).

Mendoza, B. (2005) Total Solar Irradiance and Climate Advances, Space Research, Vol. 35, pp. 882–890 (London: Elsevier Ltd).

Midwest Reliability Organization (2011) Midwest Reliability Organization (Roseville, MN: MRO) at www.midwestreliability.org/ (cited: 07/19/11).

Midwestern Accord (2010a) Midwestern Greenhouse Gas Reduction Accord (Washington, DC: MGGRA) at www.midwesternaccord.org (cited: 03/13/11).

Midwestern Accord (2010b) WCI Scope (Washington, DC: MGGRA) at www.midwesternaccord.org/./GHG-meeting-4/WCI_Update.ppt (cited: 03/13/11).

Milbourn, C. (2010) EPA to Set Modest Pace for Greenhouse Gas Standards / Agency Stresses Flexibility and Public Input in Developing Cost-effective and Protective GHG Standards for Largest Emitters (Washington, DC: Environmental Protection Agency) at http://yosemite1.epa.gov/opa/admpress.nsf/6427a6b7538955c585257359003f0230/d2f038e9daed78de8525780200568bec!OpenDocument (cited 01/08/11).

Milenio (2010) Pacto de la Ciudad de México: diez acciones contra el cambio climático (Mexico City: Milenio) at http://www.milenio.com/node/583155 (cited 12/12/10).

Miller, C. (2001) Challenges in the Application of Sciences to Global Affairs: Contingency, Trust, and Moral Order. In C. Miller and P. Edwards (eds.) Changing the Atmosphere: Expert Knowledge and Environmental Governance (Cambridge, MA: MIT Press).

Miller, C. and P. Edwards (eds.) (2001) Changing the Atmosphere: Expert Knowledge and Environmental Governance (Cambridge, MA: MIT Press).

Milne, J.E. (2008) Carbon Taxes in the United States: The Context for the Future (South Royalton, VT: Western Newspaper Publishing, Environmental Tax Policy Institute and the Vermont Law School).

Mingst, K. (2004) International Organizations: The Politics and Processes of Global Governance (Boulder, CO: Lynne Rienner Publishers, Inc.).

Ministry of the Environment—Ontario (2009) Ontario's Climate Action Plan: Cap and Trade and Industry Transition to Low Carbon Economy (Ottawa, ON: Government of Ontario) at http://www.ene.gov.on.ca/documents/ct-3_actionPlan.pdf (cited: 02/24/09).

Mitchell, R.B. (1995) Heterogeneities at Two Levels: States, Non-state Actors and International Oil Pollution. In R. Keohane and E. Ostrom (eds.) Local Commons and Global Interdependence (London: SAGE Publications).

Mittelman, J.H. (1996) Rethinking the New Regionalism in the Context of Globalization, Global Governance, Vol. 2, pp. 189–213 (Boulder, CO: Lynne Rienner Publishers).

Mol, A. (2002) The Ecological Modernization of the Global Economy, Global Environmental Politics, Vol. 2, No. 2, pp. 92–115 (Boston, MA: MIT Press).

Morales, I. (2011) Energy Trade and Security Issues at the U.S.–Mexico Border, Forum: The Future of Oil in Mexico (Mexico City: James A. Baker III Institute for Public Policy and Oxford University).

Morales, I. (2008) Post-NAFTA North America: The Challenges for Economic and Political Governance in a Changing Region (Basingstoke: Palgrave Macmillan).

Moss, R.H., A.L. Brenkert and E.L. Malone (2001) Vulnerability to Climate Change: A Quantitative Approach (Washington, DC: U.S. Department of Energy).

Mumme, S. (1999) NAFTA and Environment, Foreign Policy in Focus, Vol. 4, No. 6 (October) (Washington, DC: Interhemispheric Resource Center and Institute for Policy Studies) at http://www.fpif.org/pdf/vol4/26ifnaft.pdf (cited: 05/05/09).

Myint, T. (2007) Dynamics of Issues, Interests, and Actors in Multilayer Governance of the Rhine River Basin (Tucson, AZ: International Studies Association) at http://www.allacademic.com/meta/p_mla_apa_research_citation/1/8/0/4/9/p180492_index.html (cited: 05/17/11).

National Climate Data Center (2013) Global Climate Change Indicators: Human Influence (Asheville, NC: National Climate Data Center) at https://www.ncdc.noaa.gov/indicators/ (cited: 10/22/13).

National Energy Board—Canada (2009a) Annual Report to Parliament 2008 (Calgary: National Energy Board) at http://www.neb.gc.ca/clf-nsi/rpblctn/rprt/nnlrprt/2008/nnlrprt2008-eng.pdf (cited: 08/11/10).

National Energy Board—Canada (2009b) Canada's Energy Future: Infrastructure Changes and Challenges to 2020—An Energy Market Assessment 2009 (Calgary: National Energy Board) at http://www.neb-one.gc.ca/clf-nsi/rnrgynfmtn/nrgyrprt/nrgyftr/2009/nfrstrctrchngchllng2010/nfrstrctrchngchllng2010-eng.pdf (cited: 08/11/10).

National Energy Board—Canada (2008) Canada's Oil Sands: Opportunities and Challenges to 2015—An Update: Questions and Answers (Calgary: National Energy Board) at http://www.neb.gc.ca/clf-nsi/rnrgynfmtn/nrgyrprt/lsnd/pprtntsndchllngs20152006/qapprtntsndchllngs20152006-eng.html (cited: 08/11/10).

National Energy Policy Development Group (2001) Reliable, Affordable, and Environmentally Sound Energy for America's Future (Washington, DC: National Energy Policy Development Group) at http://www.wtrg.com/EnergyReport/National-Energy-Policy.pdf (cited: 10/30/13).

National Mining Association (2009a) A Vision for Achieving Ultra-low Emissions from Coal-fueled Electric Generation (Washington, DC: National Mining Association) at http://www.nma.org/pdf/coal_vision.pdf (cited: 02/02/11).

National Mining Association (2009b) America's Power: Coal Provides Jobs and Affordable Electricity (Washington, DC: National Mining Association) at http://www.nma.org/americas_power.asp (cited: 02/02/11).

National Renewable Energy Laboratory (2006) Interaction between Carbon Markets and Renewable Energy Markets (Golden, CO: National Renewable Energy Laboratory).

National Research Council (1995) Natural Climate Variability on Decade-to-century Time Scales, Climate Research Committee, National Research Council (Washington, DC: National Academy Press) at http://www.nap.edu/catalog/5142.html (cited: 12/09/10).

Natural Resources Canada (2009) Modernized Energy Efficiency Act Receives Royal Assent (Ottawa, ON: Natural Resources Canada) at http://www.nrcan.gc.ca/media/newcom/2009/200942-eng.php (cited: 06/14/11).

New Hampshire Business Resources (2010) Warnings Realized in RGGI Budget Raid (Manchester, NH: NHBR) at http://www.nhbr.com/businessnews/statenews/764176-257/warnings-realized-in-rggi-budget-raid.html (cited: 05/02/11).

New York State (2010) Energy (New York: Department of Environmental Conservation) at http://www.dec.ny.gov/energy (cited: 03/02/11).

New York State Energy Research and Development Authority (2010) Relative Effects of Various Factors on RGGI Electricity Sector CO2 Emissions: 2009 Compared to 2005, Draft White Paper (Red Bank, NJ: New Jersey Watchdog) at http://newjersey.watchdog.org/files/2010/11/Retrospective_Analysis_Draft_White_Paper.pdf (cited: 03/03/11).

Newcombe, K. (2008) Carbon Markets: Trends, Perspectives and US Outlook (San Francisco, CA: Goldman Sachs) at http://www2.goldmansachs.com/citizenship/environment/carbon-markets-sf.pdf (cited: 05/16/11).

Newell, P. and M. Paterson (2010) Climate Capitalism: Global Warming and the Transformation of the Global Economy (Cambridge, UK: Cambridge University Press).

North, D. (1990) Institutions, Institutional Change, and Economic Performance (Cambridge, UK: Cambridge University Press).

North America 2050 (2012) About North America 2050 (Arlington, VA: North America 2050) at http://na2050.org/ (cited: 07/06/13).

North American Agreement on Environmental Cooperation (1994) (Mexico City: North American Agreement on Environmental Cooperation).

North American Electric Reliability Corporation (2011a) About NERC (Atlanta, GA: North American Electric Reliability Corporation) at http://www.nerc.com/page.php?cid=1|7 (cited: 03/04/11).

North American Electric Reliability Corporation (2011b) Key Players: Regional Entities (Atlanta, GA: North American Electric Reliability Corporation) at http://www.nerc.com/page.php?cid=1|9|119 (cited 02/13/11).

North American Electric Reliability Corporation (2010) Map—NERC Regions (Atlanta, GA: North American Electric Reliability Corporation) at http://www.nerc.com/fileUploads/File/AboutNERC/maps/NERC_Regions_color.jpg (cited 02/13/11).

North American Energy Working Group (2012) The North America Carbon Storage Atlas (Washington, DC: North American Energy Working Group and Department of Energy—Secretaría de Medio Ambiente y Recursos Naturales—Natural Resources Canada) at http://www.netl.doe.gov/technologies/carbon_seq/refshelf/NACSA2012.pdf (cited: 06/06/13).

North American Energy Working Group (2006) North American Energy Picture II (Washington, DC: North American Energy Working Group) at http://www.pi.energy.gov/documents/NorthAmericaEnergyPictureII.pdf (cited: 12/11/10).

North American Energy Working Group (2005) North American Energy Working Group (Washington, DC: North American Energy Working Group) at http://www.libertyparkusafd.org/lp/Hamilton/Security%20and%20Prosperity%20Partnership%5CNorth%20American%20Energy%20Working%20Group.pdf (cited: 12/11/10).

Northeast Power Coordinating Council (2011) About (New York: Northeast Power Coordinating Council) at http://www.npcc.org/aboutus/about.aspx (cited: 06/07/11).

Nye, J.S. and J. Donahue (eds.) (2000) Governance in a Globalizing World (Washington, DC: Brookings Institution Press).

O'Connor, P.A. (2010) Energy Transitions, The Pardee Papers, No. 12 (November) (Boston, MA: University of Boston).

O'Donnell, G. (1993) Delegative Democracy? Working Paper No. 192 (Notre Dame, IN: Kellogg Institute for International Studies) at http://www.yorku.ca/sbohn/pols4555/Notes/2010/Week%208_O'Donnell_delegative%20democracy.pdf (cited: 12/11/10).

Oficina de Presidencia—Mexico (2008) Mexico to Submit Proposal to Create Green Fund to G8 Environment Ministers (Mexico City: Oficina de Presidencia).

Ohmae, Kenichi (1995) The End of the Nation State (New York: Free Press).

Organization for Economic Cooperation and Development (2009) The Economics of Climate Change Mitigation (Paris: OECD).

Ostrom, E. (1995) Constituting the Social Capital and Collective Action. In R. Keohane and E. Ostrom (eds.) Local Commons and Global Interdependence (London: SAGE Publications).

Ostrom, E. (1992) Community and the Endogenous Solution of Commons Problems, Journal of Theoretical Politics, Vol. 4, No. 3 (July), pp. 343–352 (London: SAGE Publications).

Ostrom, E. (1990) Governing the Commons: The Evolution of Institutions for Collective Action (New York: Cambridge University Press).

Oye, K.A. and M.H. James (1995) Self-interest and Environmental Management. In R. Keohane and E. Ostrom (eds.) Local Commons and Global Interdependence (London: SAGE Publications).

Pacific Carbon Trust (2011) Pacific Carbon Trust (Vancouver, BC: Pacific Carbon Trust) at pacificcarbontrust.com (cited: 07/07/11).

Paehlke, R. (2008) Some like it Cold: The Politics of Climate Change in Canada (Toronto, ON: Between the Lines).

Parliament of Canada (2012) Bill C-224 (Ottawa, ON: Parliament of Canada) at http://www.parl.gc.ca/LEGISInfo/BillDetails.aspx?Mode=1&billId=5087216 &Language=E (cited 05/06/13).

Parliament of Canada (2010) Bill C-311 (Ottawa, ON: Parliament of Canada) athttp://www2.parl.gc.ca/HousePublications/Publication.aspx?Docid=3662654& File=9 (cited 01/10/11).

Parliament of Canada (2007a) Bill C-30: Canada's Clean Air and Climate Change Act (LS-539E) (Ottawa, ON: Parliament of Canada) at http://www2.parl. gc.ca/Sites/LOP/LegislativeSummaries/Bills_ls.asp?lang=E&ls=c30&source =library_prb&Parl=39&Ses=1 (cited: 10/26/10).

Parliament of Canada (2007b) Kyoto Protocol Implementation Act (Ottawa, ON: Parliament of Canada) at http://laws.justice.gc.ca/en/showdoc/cs (cited 10/26/10).

Partido Verde Ecologista de Mexico (2011) Presenta la Dip. Gabriela Medrano Iniciativa de Ley de Cambio Climático (Mexico City: Partido Verde Ecologista de Mexico) at http://www.partidoverde.org.mx/pvem/2011/06/presenta-la-dip-gabriela-medrano-iniciativa-de-ley-de-cambio-climatico/ (cited: 06/15/11).

Paul, T.V. (ed.) (2012) International Relations Theory and Regional Transformation (Cambridge, UK: Cambridge University Press).

Pembina Foundation (2008) Greenlearning (Drayton Valley, AB: Pembina Foundation) at http://www.greenlearning.ca/climate/policy/canadian-policy (cited: 05/04/08).

Perkins Coie & Fasken Martineau (2008) Government Financial Incentives to Support Greenhouse Gas Reductions (Seattle, WA: Perkins Coie & Fasken Martineau) at http://www.perkinscoie.com/news/pubs_detail.aspx?publication =1792 (cited: 07/17/10).

Pew Center on Global Climate Change (2010a) Climate Action—Table (Arlington, VA: Pew Center on Global Climate Change) at http://www.pewclimate.org/ states-regions (cited: 09/10/11).

Pew Center on Global Climate Change (2010b) Regional Greenhouse Gas Initiative (RGGI) (Arlington, VA: Pew Center on Global Climate Change) at http://www. pewclimate.org/what_s_being_done/in_the_states/rggi (cited: 07/07/11).

Pew Center on Global Climate Change (2009a) Climate Change 101: State Action (Arlington, VA: Pew Center on Global Climate Change) at http://www. pewclimate.org/docUploads/Climate101-State-Jan09.pdf (cited 02/18/10).

Pew Center on Global Climate Change (2009b) Myths about the Waxman–Markey Clean Energy Bill (Arlington, VA: Pew Center on Global Climate Change) at http://www.pewclimate.org/acesa/eight-myths/June2009 (cited: 09/10/11).

Pew Center on Global Climate Change (2009c) Waxman–Markey Short Summary (Arlington, VA: Pew Center on Global Climate Change) at http://www. pewclimate.org/federal/analysis/congress/111/acesa/waxman-markey-short-summary (cited: 09/10/11).

Pierre, J. (coord.) (2000) Debating Governance: Authority, Steering and Democracy (Oxford: Oxford University Press).

Pizer, W. (2007) Practical Global Climate Change Policy. In J.E. Aldy and R.N. Stavins (eds.) Architectures for Agreement: Addressing Global Climate Change in the Post-Kyoto World (Cambridge, UK: Cambridge University Press).

Plymouth Marine Laboratory (2013) Carbon Capture and Storage (Plymouth, UK: Plymouth Marine Laboratory) at http://www.pml.ac.uk/science_areas/todays_models/carbon_capture_and_storage.aspx (cited: 06/10/13).

Poloncarz, K., M.S. Balster and B.B. Carrier (2013) May 2013 California Cap-and-trade Auction Shows Strengthening Carbon Market (London: Globe Business Publishing, LTD, Paul Hastings, LLP, and Lexology) at http://www.lexology.com/library/detail.aspx?g=ca21c794-08aa-40e8-9bd5-9e84f9bd6bdb (cited: 07/02/13).

Potomac Economics (2010) Market Monitor Report for Auction 10 (Fairfax, VA: Potomac Economics) at http://www.rggi.org/docs/Auction_10_Market_Monitor_Report.pdf (cited: 03/03/11).

Prado, V. (2007) Energy Infrastructure in the Western Hemisphere. In S. Weintraub, A. Hester and V. Prado (eds.) Energy Cooperation in the Western Hemisphere: Benefits and Impediments (Washington, DC: Center for Strategic and International Studies Press).

Press Republican (2009) Montreal–New York City High-speed Rail a Priority for Quebec (New York: Press Republican) at http://pressrepublican.com/archive/x546312957/Montreal-New-York-City-high-speed-rail-a-priority-for-Quebec (cited: 01/23/11).

Puyana, A. (2006) Fuerzas que modelan la política energética mexicana. In I. Rousseau, ¿Hacia la integración de los mercados petroleros en América? (Mexico City: El Colegio de México).

Quadri, G. (2007) Calentamiento global, bienes públicos y mercado de carbono, Foreign Affairs en Español (Julio–Diciembre) (Mexico City: Instituto Tecnológico Autónomo de México).

Rabe, B. (2010) Building on Sub-federal Climate Strategies: The Challenge of Regionalism, Working Paper, Workshop-designing Integration: Regional Governance on Climate Change in North America (September) (Waterloo, ON: University of Waterloo, Wilfrid Laurier University and Instituto Tecnológico de Monterrey).

Ramos, J.M. (2002) Gobiernos locales y la cooperación transfronteriza entre México-Estados Unidos, Espiral, Estudios sobre Estado y Sociedad 9, no. 25 (Septiembre–Diciembre), pp. 107–140 (Guadalajara: Universidad de Guadalajara) at http://redalyc.uaemex.mx/redalyc/pCity/138/13802504.pdf (cited: 11/23/08).

Raustiala, K. (1999) States, NGOs, and Environmental International Institutions, International Studies Quarterly, Vol. 41, pp. 719–741 (London: Wiley-Blackwell).

Ray, A. and S. Phoha (2005) Homeostasis and Homeorhesis: Sustaining Order and Normalcy in Human-engineered Complex Systems (University Park, PA: Pennsylvania State University) at http://www.mne.psu.edu/ray/interdisciplinaryresearch.pdf (cited: 06/12/13).

Regional Greenhouse Gas Initiative (2013a) Database—Auctions Results (New York: RGGI) at http://www.rggi.org/market/co2_auctions/results (cited: 06/26/13).

Regional Greenhouse Gas Initiative (2013b) Model Rule (New York: RGGI) at http://www.rggi.org/docs/ProgramReview/_FinalProgramReviewMaterials/Model_Rule_FINAL.pdf (cited: 06/26/13).

Regional Greenhouse Gas Initiative (2011a) Fact Sheet: RGGI Offsets (New York: RGGI) at http://www.rggi.org/docs/RGGI_Offsets_in_Brief.pdf (cited: 12/12/10).

Regional Greenhouse Gas Initiative (2011b) RGGI, Inc. (New York: RGGI) at http://www.rggi.org/rggi (cited: 12/12/10).

Regional Greenhouse Gas Initiative (2010a) CO2 Auctions, Tracking & Offsets (New York: RGGI) at http://www.rggi.org/market (cited: 02/22/11).

Regional Greenhouse Gas Initiative (2010b) Fact Sheet: RGGI CO2 Allowance Tracking System (RGGI COATS) (New York: RGGI) at http://www.rggi.org/docs/RGGI_COATS_in_Brief.pdf (cited: 02/22/11).

Regional Greenhouse Gas Initiative (2010c) RGGI COATS Data (New York: RGGI) at www.rggi.org/market/tracking/data (cited: 02/22/11).

Rhodes, R.A.W. (1996) The New Governance: Governing without Government, Political Studies, Vol. 44, No. 4, pp. 652–657 (Hoboken, NJ: John Wiley & Sons, Inc.).

Rhodes, R.A.W. and D. Marsh (1992) New Directions in the Study of Policy Networks, European Journal of Political Research, Vol. 21, Nos. 1–2, pp. 181–205 (London: Wiley-Blackwell).

Richardson, N. (2012) Playing without ACES: Offsets and the Limits of Flexibility under Clean Air Climate Policy, Environmental Law, Vol. 42, No. 3 (Summer), pp. 735–783.

Rincon, P. (2013) Obama Lays Out Climate Action Plan (London: BBC News) at http://www.bbc.co.uk/news/science-environment-23032890 (cited: 06/26/13).

Risse, T. (2011) Governance without a State? Policies and Politics in Areas of Limited Statehood (New York: Columbia University Press).

Rosenau, J.N. (1997) Along the Domestic–Foreign Frontier: Exploring Governance in a Turbulent World (Cambridge, UK: Cambridge University Press).

Rosenau, J.N. (1992) Governance, Order, and Change in World Politics. In J. Rosenau and E. Czempiel (eds.) Governance without Government: Order and Change in World Politics (Cambridge, UK: Cambridge University Press).

Rosenau, J.N. (1990) Turbulence in World Politics: A Theory of Change and Continuity (Princeton, NJ: Princeton University Press).

Rosenau, J.N. and E. Czempiel (eds.) (1992) Governance without Government: Order and Change in World Politics (Cambridge, UK: Cambridge University Press).

Rosenbaum, W.A. (1987) Energy, Politics, and Public Policy (Washington, DC: Congressional Quarterly, Inc.).

Rousseau, I. (2010) PEMEX y la política petrolera: los retos hacia el futuro. In J. L. Méndez (ed.) Los grandes problemas de México: XIII Políticas Públicas, Edición Abreviada (Mexico City: El Colegio de México).

Rousseau, I. (2006) ¿Hacia la integración de los mercados petroleros en América? (Mexico City: El Colegio de México).

Rowlands, I. (2008) Integrating Climate Policy and Energy Policy. In S. Bernstein, J. Brunnée, D. Duff and A. Green (eds.) A Globally Integrated Climate Policy for Canada (Toronto, ON: University of Toronto Press).

Salomon, J. (2000) Science, Technology and Democracy, Minerva, Vol. 38, pp. 33–51 (Netherlands: Kluwer Academic Publishers).

Samuelsohn, D. (2010) Boxer, Kerry Set to Introduce Climate Bill in Senate (New York: The New York Times) at http://www.nytimes.com/cwire/2009/09/28/28climatewire-boxer-kerry-set-to-introduce-climate-bill-in-43844.html (cited: 03/02/11).

Samuelson, P. (1954) The Pure Theory of Public Expenditure, The Review of Economics and Statistics, Vol. 36, No. 4 (November), pp. 387–389 (Boston, MA: MIT Press).

Sbragia, Alberta (2010) Multi-level Governance and Comparative Regionalism. In H. Enderlein, S. Wälti and M. Zürn (eds.) Handbook on Multi-level Governance (Cheltenham, UK: Edward Elgar Publishing).

Schelling, T. (1993) Some Economics of Global Warming. In R. Dorfman and N. Dorfman (eds.) Economics of the Environment, Selected Readings, Third Edition (New York: W.W. Norton and Co.).

Scherer, A.G. and G. Palazzo (eds.) (2008) Handbook of Research on Global Corporate Citizenship (Cheltenham, UK and Northampton, MA: Edward Elgar Publishing).

Schlesinger, W.H. (2000) Biogeoquímica, un análisis global (Barcelona: Ariel Ciencia).

Scholte, J.A. (2000) Globalization: A Critical Introduction (New York: St. Martin's Press).

Schott, J. (2008) NAFTA and the Environment (Washington, DC: Peterson Institute for International Economics) at http://www.cec.org/files/pdf/ABOUTUS/Schott-NAFTA%20and%20the%20Environment_en.pdf (cited: 11/10/10).

Schultz, S. (2012) British Columbia Rethinks Its Pioneering Carbon Tax, National Geographic News (London: National Geographic) at http://news.nationalgeographic.co.uk/news/energy/2012/05/120503-british-columbia-reviews-carbon-tax/ (cited: 07/01/13).

Schwanen, D. (2006) Canada and the Kyoto Protocol: When Reality Sets In. In A. Cooper and D. Rowlands (eds.) Canada among Nations 2006 (Montreal, QC and Kingston, ON: McGill-Queen's University Press).

Seeberg-Elverfeldt, C. (2010) Carbon Finance Possibilities for Agriculture, Forestry and Other Land Use Projects in a Smallholder Context (Rome: Food and Agriculture Organization of the United Nations) at http://www.fao.org/docrep/012/i1632e/i1632e02.pdf (cited: 05/16/11).

Sending, O.J. and I.B. Neumann (2007) Governance to Governmentality: Analyzing NGOs, States, and Power, International Studies Quarterly, Vol. 50, No. 3, pp. 653–672 (London: Wiley-Blackwell).

Secretaría de Energía (2012a) Exportaciones de petrolíferos (Mexico City: Secretaría de Energía).

Secretaría de Energía (2012b) Prospectiva de Petróleo Crudo (Mexico City: Secretaría de Energía).

Secretaría de Energía (2011a) Estrategia Nacional para la Transición Energética y el Aprovechamiento Sustentable de la Energía (Mexico City: Secretaría de Energía).

Secretaría de Energía (2011b) Quinto Informe de Labores (Mexico City: Secretaría de Energía).

Secretaría de Energía (2007a) Prospectivas del Mercado de Gas Natural 2007–2016 (Mexico City: Secretaría de Energía) at http://energia.gob.mx/webSener/res/PE_y_DT/pub/Prospectiva%20Gas%20Natural%202007-2016%20FINAS.pdf (03/02/11).

Secretaría de Energía (2007b) Prospectivas del Mercado de Petróleo Crudo 2007–2016 (Mexico City: Secretaría de Energía) at http://www.sener.gob.mx/webSener/res/PE_y_DT/pub/Prospectiva%20Petroleo%20Crudo%20Finas.pdf (03/02/11).

Secretaría de Energía (2007c) Prospectivas del Sector Eléctrico 2007–2016 (Mexico City: Secretaría de Energía) at http://www.energia.gob.mx/webSener/res/PE_y_DT/pub/Prospectiva%20Sector%20Electrico%20FINAS.pdf (03/02/11).

Secretaría de Medio Ambiente y Recursos Naturales (2009a) Comisión Intersecretarial de Cambio Climático (Mexico City: Secretaría de Medio Ambiente y Recursos Naturales) at http://www.semarnat.gob.mx/TEMAS/CAMBIOCLIMATICO/Paginas/cicc.aspx (cited: 11/11/10).

Secretaría de Medio Ambiente y Recursos Naturales (2009b) Programa Especial de Cambio Climático 2009–2012 (Mexico City: Secretaría de Medio Ambiente y Recursos Naturales) at http://www.semarnat.gob.mx/queessemarnat/politica_ambiental/cambioclimatico/Documents/pecc/090828_PECC.Capitulos_DOF.pdf (cited: 11/11/10).

Sheberle, D. (2004) Devolution. In R. Durant, D. Fiorino and R. O'Leary (eds.) Environmental Governance Reconsidered: Challenges, Choices and Opportunities (Cambridge, MA: MIT Press).

Sheppard, R. (2006) The Difference between Kyoto and the New U.S.-led Climate Pact that Stephen Harper is Suddenly Keen to Join (Toronto, ON: Canadian Broadcasting Corporation) at http://www.cbc.ca/news/background/realitycheck/sheppard/20060519.html (consulted: 05/05/2009).

Sikkink, K. (2003) La dimensión transnacional de los movimientos sociales. In E. Jelin (ed.) Más allá de la nación: las escalas múltiples de los movimientos sociales (Buenos Aires: Libros del Zorzal).

Simeon, R., G. Hoberg and K. Banting (1997) Degrees of Freedom: Canada and the United States in a Changing World (Montreal, QC and Kingston, ON: McGill-Queen's University Press).

Simpson, J., M. Jaccard and N. Rivers (2007) Hot Air: Meeting Canada's Climate Change Challenge (Toronto, ON: Emblem—McClelland & Stewart Corporation).

Sindzingre, A. (2006) The Relevance of the Concepts of Formality and Informality: A Theoretical Appraisal. In B. Guha-Knasnobis, R. Kanbur and E. Ostrom, Linking the Formal and Informal Economy: Concepts and Policies (Oxford, UK: Oxford University Press, UNU-WIDER Studies in Development Economics, and EGDI) at http://www.oecd.org/dataoecd/52/25/37791314.pdf (cited: 07/07/11).

Slaughter, A.M. and T.N. Hale (2010) Transgovernmental Networks and Multi-level Governance. In H. Enderlein, S. Wälti and M. Zürn (eds.) Handbook on Multi-level Governance (Cheltenham, UK: Edward Elgar Publishing).

Sloan, W.M., P. Hsiao, M. Steel and M.A. Jennings (2013) United States: Picking Up Steam? CARB Reports that California's Second Carbon Auction Exceeds Price Expectations (San Francisco, CA: Morrison & Foerster—Mondaq) at http://www.mondaq.com/unitedstates/x/224708/Clean+Air+Pollution/Picking+Up+Steam+CARB+Reports+That+Californias+Second+Carbon+Auction+Exceeds+Price+Expectations (cited: 07/02/13).

Smith, H. (1998) Canadian Federalism and Environmental Policy Making: The Case of Climate Change, Working Paper No. 5 (Kingston, ON: Institute for Intergovernmental Relations, Queen's University Press).

Smith, V.K., K.A. Shwabe and C. Mansfield (1997) Does Nature Limit Environmental Federalism?, Duke Economics Working Paper 97-01, (Washington, DC: Resources for the Future) at http://ssrn.com/abstract=56042 (cited 05/02/09).

Snidal, D (1995) The Politics of Scope: Endogenous Actors, Heterogeneity and Institutions. In R. Keohane and E. Ostrom (eds.) Local Commons and Global Interdependence (London: SAGE Publications).

Söderbaum, F. (2004) Introduction: Theories of New Regionalism. In F. Söderbaum and T.M. Shaw (eds.) Theories of New Regionalism (Basingstoke: Palgrave Macmillan).

Söderbaum, F. and T.M. Shaw (eds.) (2004) Theories of New Regionalism (Basingstoke: Palgrave Macmillan).

Soldatos, P. and H. Michelmann (1990) Federalism and International Relations: The Role of Subnational Units (Oxford, UK: Clarendon Press).

State Initiatives for Renewables and Efficiency (2009) Rules, Regulations & Policies for Renewable Energy: Database (Raleigh, NC: DSIRE—North Carolina State University) at http://www.dsireusa.org/summarytables/rrpre.cfm (cited: 02/01/11).

State of California (2013) U.S. Per Capita Electricity Use by State in 2010 (Sacramento, CA: State of California) at http://energyalmanac.ca.gov/electricity/us_per_capita_electricity-2010.html (cited: 07/04/13).

Statistics Canada (2011) Energy Statistics Handbook: Second Quarter 2011 (Ottawa, ON: Statistics Canada) at http://www.statcan.gc.ca/pub/57-601-x/57-601-x2011002-eng.pdf (cited: 07/02/13).

Statistics Canada (2010) Canada's Estimated Population (Ottawa, ON: Statistics Canada) at http://www.statcan.gc.ca/daily-quotidien/100628/dq100628a-eng.htm (cited: 02/15/11).

Statistics Canada (2009) Canada's Real Gross Domestic Product (Ottawa, ON: Statistics Canada) at http://www40.statcan.ca/l01/cst01/econ50-eng.htm (cited: 11/04/10).

Stavins, R.N. (2012) Low Prices a Problem? Making Sense of Misleading Talk about Cap-and-trade in Europe and the USA: An Economic View of the Environment (Cambridge, MA: Kennedy School and The President and Fellows of Harvard College).

Sterling-Folker, Jennifer (2006) Making Sense of International Relations Theory (Boulder, CO: Lynne Rienner Publishers).

Sterman, J.D. and L.B. Sweeney (2002) Cloudy Skies: Assessing Public Understanding of Global Warming, System Dynamics Review, Vol. 18, No. 2, pp. 207–240 (London: Wiley-Blackwell).

Stern, N. (2007) The Economics of Climate Change: The Stern Review (Cambridge, UK: Cambridge University Press).

Stevenson, G. (1989) Federalism in Canada (Toronto, ON: McClelland & Stewart, Inc.).

Stikeman Elliott (2013) Federal Environment Minister Expects New Oil and Gas Regulations by Mid-Year (Montreal, QC: Stikeman Elliott—Canadian Energy Law) at http://canadianenergylaw.com/articles/climate-change/ (cited 06/05/13).

Stockholm Environment Institute (2010) Midwestern Greenhouse Gas Reduction Accord: Overview (Stockholm: Stockholm Environment Institute) at http://www.co2offsetresearch.org/policy/MGGRA.html (cited 02/03/11).

Stubbs, R. and S. Eaton (2006) State Responses to Globalization. In R. Stubbs and G. Underhill (eds.) Political Economy and the Changing Global Order (Oxford, UK: Oxford University Press).

Stubbs, R. and G. Underhill (eds.) (2006) Political Economy and the Changing Global Order (Oxford, UK: Oxford University Press).

Sustainable City Network (2010) ICLEI USA Statement on Final Agreement of Cancun COP16 Negotiations (Dubuque, IA: Sustainable City Network, Inc.) at http://www.sustainablecitynetwork.com/topic_channels/policy/article_2d136926-084b-11e0-bbf7-0017a4a78c22.html (cited 12/20/10).

Tangen, K. and H. Hasselknippe (2005) Converging Markets, International Environmental Agreements: Politics, Law and Economics, Vol. 5, No. 1, pp. 47–64 (Dordrecht: Springer).

Terizakis, G. (2010) Sustainability and Environmental Policy Integration: Local Level Sustainability through Knowledge Involvement and New Governance Arrangements. In A. Goria, A. Sgobbi and I. von Homeyer (eds.) Governance for the Environment: A Comparative Analysis of Environmental Policy Integration (Cheltenham, UK: Edward Elgar Publishing).

The Canadian Bioenergy Association (2005) Renewable Portfolio Standard (RPS) & Other Incentives: Harmonization Opportunities in Canada (Ottawa, ON: The Canadian Bioenergy Association) at http://www.canbio.ca/documents/publications/Renewable_Harmonize_Final_Feb_16_2005.pdf (cited 12/16/10).

The Climate Action Tracker (2011) Durban Agreements a Step Towards a Global Agreement, but Risk of Exceeding 3°C-Warming Remains—Scientists (Durban: The Climate Action Tracker) at http://climateactiontracker.org/news/116/Durban-Agreements-a-step-towards-a-global-agreement-but-risk-of-exceeding-3C-warming-remains-scientists.html (cited: 12/14/12).

The Economist (2009) The Greening of Canada? Stephen Harper Looks to Washington, not Alberta (New York: The Economist Newspaper Limited) at http://www.economist.com/daily/columns/greenview/displaystory.cfm?story_id=13167651&fsrc=nwl (cited: 02/23/09).

The Energy Roundtable (2010) Information on Canadian Energy Opportunities (Ottawa, ON: The Energy Roundtable) at http://www.energyroundtable.org/energy_opp.php (cited 07/30/10).

The Environics Institute (2011) Focus Canada 2011 (Toronto, ON: The Environics Institute) at http://www.environicsinstitute.org/uploads/institute-projects/pdf-focuscanada-2011-final.pdf (cited: 06/17/13).

The Transportation & Climate Initiative of the North-east and Mid-Atlantic States (2010) Overview of the Regional Greenhouse Gas Initiative (Washington, DC: Georgetown University Press) at http://www.georgetownclimate.org/transportation/files/TCI-Overview%20of%20RGGI.PDF (cited: 04/09/11).

The White House (2012) State of the Union Address (Washington, DC: The White House) at http://www.whitehouse.gov/state-of-the-union-2012 (cited: 07/04/12).

The White House (2011) State of the Union Address (Washington, DC: The White House) at http://www.whitehouse.gov/state-of-the-union-2011 (cited: 03/04/12).

The White House (2010) State of the Union Address (Washington, DC: The White House) at http://www.whitehouse.gov/the-press-office/remarks-president-state-union-address (cited: 03/04/11).

The White House (2009) North American Leaders' Declaration on Climate Change and Clean Energy (Washington, DC: The White House) at http://www.whitehouse.gov/the_press_office/North-American-Leaders-Declaration-on-Climate-Change-and-Clean-Energy/ (cited: 02/13/11)

Thomas, D.M. and B.B. Torrey (eds.) (2008) Canada and the United States: Differences that Count, Third Edition (Peterborough, ON: Broadview Press).

Thompson, J.D. (1967) Organizations in Action (New York: McGraw-Hill).

Three-Regions Offsets Working Group (2010) Ensuring Offset Quality: Design and Implementation Criteria for a High-quality Offset Program (Washington, DC: RGGI) at http://www.rggi.org/docs/3_Regions_Offsets_Announcement_05_17_10.pdf (cited: 06/28/11).

Toledo, V. and P. Moguel (1992) Ecología, geografía y producción rural: el problema de la concepción de la naturaleza, Revista Relaciones, Vol. 13, No. 50, pp. 7–22 (Zamora: Colegio de Michoacán) at www.colmich.edu.mx/relaciones/indices/articulos/articuloDetalle.asp?id=436-4k (cited: 10/15/08).

Torres, B. (2009) North American Transnational Environmental Actors and Mexico. In J. Ayres and L. Macdonald (eds.) Contentious Politics in North America (New York: Palgrave Macmillan).

Trisko, E.M. (2010) Statement on Behalf of the United Mine Workers of America, AFL-CIO Before the Committee on Energy and Commerce Energy and Environment Subcommittee U.S. House of Representatives April 23, 2009, "The American Clean Energy and Security Act of 2009" (Berkeley Springs, WV: Union for Jobs and Environment) at http://www.ujae.org/NewsRes2009/UMWA%20Waxman%20Markey%20Statement%20042309.pdf (cited: 02/11/11).

Tuerk, A., M. Mehling, C. Flachsland and W. Sterk (2009) Linking Emissions Trading Schemes, Climate Policy, Vol. 9, No. 4, pp. 341–357 (London: Earthscan/Climate Strategies).

Turnbull, Sh. (2002) Corporate Governance: Its Scope, Concerns and Theories, Corporate Governance: An International Review, Vol. 5, No. 4, pp. 180–205 (London: Wiley-Blackwell).

United Nations Environmental Program (2012a) Global Environmental Outlook GEO5 (Nairobi: United Nations Environmental Program) at http://www.unep.org/geo/pdfs/geo5/GEO5_report_full_en.pdf (cited: 07/09/13).

United Nations Environmental Program (2012b) The Emissions Gap Report 2012: A UNEP Synthesis Report (Nairobi: United Nations Environmental Program) at http://www.unep.org/publications/ebooks/emissionsgap2012/ (cited: 07/08/13).

United Nations Environmental Program (2009) Emission Reduction Pledges (Nairobi: United Nations Environmental Program) at http://www.unep.org/climatepledges/ (cited: 12/15/10).

United Nations Environmental Program (2007) Global Environmental Outlook GEO4 Environment for Development (Nairobi: United Nations Environmental Program) at http://www.unep.org/geo/geo4/report/GEO-4_Report_Full_en.pdf (cited: 12/18/09).

United Nations Environmental Program (1972) Declaration of the United Nations Conference on the Human Environment (Nairobi: United Nations Environmental Program) at http://www.unep.org/Documents.Multilingual/Default.asp?DocumentID=97&ArticleID=1503 (cited: 12/18/09).

United Nations Environmental Program and GRID-Arendal (2002) The Greenhouse Gas Effect (Arendal, Norway: UNEP/GRIDA) at http://maps.grida.no/go/graphic/greenhouse-effect (cited: 04/27/11).

United Nations Framework Convention on Climate Change (2013a) Greenhouse Gas Inventory Data (Bonn: United Nations Framework Convention on Climate Change) at http://unfccc.int/di/DetailedByCategory/Event.do?event=go (cited: 07/01/13).

United Nations Framework Convention on Climate Change (2013b) Meetings (Bonn: United Nations Framework Convention on Climate Change) at http://unfccc.int/meetings/items/6240.php (07/01/13).

United Nations Framework Convention on Climate Change (2012) Kyoto Protocol Status of Ratification (Bonn: United Nations Framework Convention on Climate Change) at http://unfccc.int/kyoto_protocol/status_of_ratification/items/2613.php (cited: 07/09/13).

United Nations Framework Convention on Climate Change (2011a) Essential Background (Bonn: United Nations Framework Convention on Climate Change) at http://unfccc.int/essential_background/convention/convention_bodies/items/2629.php (cited: 07/23/11).

United Nations Framework Convention on Climate Change (2011b) Kyoto Protocol Bodies (Bonn: United Nations Framework Convention on Climate Change) at (http://unfccc.int/kyoto_protocol/kyoto_protocol_bodies/items/2772.php) (cited: 07/23/11).

United Nations Framework Convention on Climate Change (2011c) Kyoto Protocol Bodies: SBSTA and SBI (Bonn: United Nations Framework Convention on Climate Change) at http://unfccc.int/kyoto_protocol/kyoto_protocol_bodies/items/2772.php (cited: 07/23/11).

United Nations Framework Convention on Climate Change (2011d) Kyoto Protocol Status of Ratification (Bonn: United Nations Framework Convention on Climate Change) at http://unfccc.int/kyoto_protocol/status_of_ratification/items/2613.php (cited: 07/23/11).

United Nations Framework Convention on Climate Change (2011e) Mechanisms (Bonn: United Nations Framework Convention on Climate Change) at http://unfccc.int/kyoto_protocol/mechanisms/items/1673.php (cited: 07/23/11).

United Nations Framework Convention on Climate Change (2011f) Meetings Archive (Bonn: United Nations Framework Convention on Climate Change) at http://unfccc.int/meetings/archive/items/2749.php (cited: 07/23/11).

United Nations Framework Convention on Climate Change (2011g) Parties and Observers (Bonn: United Nations Framework Convention on Climate Change) at http://unfccc.int/parties_and_observers/parties/items/2352.php (cited: 07/23/11).

United Nations Framework Convention on Climate Change (2011h) UNFCCC Database (Bonn: United Nations Framework Convention on Climate Change) at http://unfccc.int/ghg_data/ghg_data_unfccc/items/4146.php (cited: 07/23/11).

United Nations Framework Convention on Climate Change (2010a) Clean Development Mechanism Statistics (Bonn: United Nations Framework Convention on Climate Change) at http://cdm.unfccc.int/Statistics/index.html (cited: 04/03/11).

United Nations Framework Convention on Climate Change (2010b) UN Climate Change Conference in Cancun Delivers Balanced Package of Decisions, Restores Faith in Multilateral Process: Press Release (Cancun: United Nations Framework Convention on Climate Change) at http://unfccc.int/files/press/news_room/press_releases_and_advisories/application/pdf/pr_20101211_cop16_closing.pdf (cited: 02/03/11).

United Nations Framework Convention on Climate Change (2009a) COP-15 (Bonn: United Nations Framework Convention on Climate Change) at http://unfccc.int/meetings/cop_15/items/5257.php (cited: 04/03/10).

United Nations Framework Convention on Climate Change (2009b) The Relationship between Efforts to Protect the Ozone Layer and to Mitigate Climate Change: Ways and Means to Limit Emissions of HFCs and PFCs (Bonn: United Nations Framework Convention on Climate Change) at http://unfccc.int/cop6/issues/102.html (cited: 04/03/10).

United Nations Framework Convention on Climate Change (2007a) Conference of Parties (COP-13) United Nations Framework Convention for Climate Change, Bali (December 10–16) (Bonn: United Nations Framework Convention on Climate Change) at http://www.ipcc.ch/ (cited: 08/20/09).

United Nations Framework Convention on Climate Change (2007b) The Montreal Protocol and the Kyoto Protocol Mutually Supportive Say Top UN Officials (Bonn: United Nations Framework Convention on Climate Change) at http://unfccc.int/files/press/news_room/press_releases_and_advisories/application/pdf/070917_hcfc_pressrel.pdf (cited: 08/20/09).

United Nations Framework Convention on Climate Change (2007c) UNFCCC Press Release (Bonn: United Nations Framework Convention on Climate Change) at http://unfccc.int/files/press/news_room/press_releases_and_advisories/application/pdf/070917_hcfc_pressrel.pdf (cited: 08/20/09).

United Nations Framework Convention on Climate Change (1998) Kyoto Protocol to the United Nations Framework for Climate Change (Bonn: United Nations Framework Convention on Climate Change) at http://unfccc.int/resource/docs/convkp/kpeng.pdf (cited: 08/20/09).

United Nations Framework Convention on Climate Change (1992) United Nations Framework Convention on Climate Change (Bonn: United Nations

Framework Convention on Climate Change) at http://unfccc.int/resource/docs/convkp/conveng.pdf (cited: 08/20/09).

United Nations Population Division (2011) World Population to Reach 10 billion by 2100 if Fertility in All Countries Converges to Replacement Level (New York: United Nations Population Division) at http://esa.un.org/unpd/wpp/Other-Information/Press_Release_WPP2010.pdf (cited: 06/23/11).

United States Global Research Program (1997) Scientific Assessment (Washington, DC: United States Global Research Program) at http://www.globalchange.gov/publications/reports/scientific-assessments/us-impacts/regional-climate-change-impacts/southwest (cited: 04/20/11).

Van Kooten, G.C. (2004) Economic Aspects of Climate Change. In H. Coward and A. Weaver (eds.) Hard Choices: Climate Change in Canada (Waterloo, ON: Wilfrid Laurier University Press).

Van Nijnatten, D. (2008a) Canadian Environmental Policy in North American Context: Manoeuvring toward Mediocrity. In R. Bothwell and J. Daudelin (eds.) Canada among Nations: 100 Years of Canadian Foreign Policy (Montreal, QC and Kingston, ON: McGill-Queen's University Press).

Van Nijnatten, D. (2008b) Environmental Policy in Canada and the United States: Climate Change and Continuing Distinctiveness. In D.M. Thomas and B.B. Torrey (eds.) Canada and the United States: Differences that Count, Third Edition (Peterborough, ON: Broadview Press).

Van Nijnatten, D. and G.W. Boychuk (2006) Comparative State and Provincial Public Policy. In C. Dunn (ed.) Provinces: Canadian Provincial Politics (Peterborough, ON: Broadview Press).

Vargas, R. and A. Hickman (2010) La integración energética en América del Norte y la reforma energética en México (Mexico City: Centro de Investigaciones sobre América del Norte—UNAM).

Vargas, R. and J.L. Valdés (2007) Dos modelos de integración energética. América del Norte y América del Sur (Mexico City: Centro de Investigaciones sobre América del Norte-UNAM).

Verdansky, V. (1926) The Biosphere, trans. David B. Langmuir, ed. Mark A.S. McMenamin (1998) (New York: Copernicus Springer-Verlag).

Victor, D. (2007) Fragmented Carbon Markets and Reluctant Nations: Implications for the Design of Effective Architectures. In J.E. Aldy and R.N. Stavins (eds.) Architectures for Agreement: Addressing Global Climate Change in the Post-Kyoto World (Cambridge, UK: Cambridge University Press).

Villareal, M.A. and J.E. Lake (2009) Prosperity and Security of North America: An Overview of Selected Issues—RS22701 (Washington, DC: Congressional Research Services) at http://www.fas.org/sgp/crs/row/RS22701.pdf (cited: 03/03/11).

Viotti, P. R. and M. V. Kauppi (2010) International Relations Theory, Fourth Edition (New York: Pearson Education Inc.).

Wall Street Journal (2009) Henry Waxman Has a Plan … For Your Living Room, Showerhead, Jacuzzi … (April) (New York: Wall Street Journal) at http://online.wsj.com/article/SB123933057062907775.html (cited: 12/18/10).

Walker, D. (2010) States Blaze the Trail to a Clean Energy Future (New York: Environmental Defense Fund) at http://blogs.eCity.org/californiadream/2010/07/27/states-blaze-the-trail-to-a-clean-energy-future/ (cited: 02/15/11).

Wallace, H. and W. Wallace (2000) Policy-making in the European Union, Fourth Edition (Oxford: Oxford University Press).

Wälti, Sonja (2010) Multi-level Environmental Governance. In H. Enderlein, S. Wälti and M. Zürn (eds.) Handbook on Multi-level Governance (Cheltenham, UK: Edward Elgar Publishing).

Watts, R. (1999) Comparing Federal Systems, Second Edition (Kingston, ON: Queen's University Press).

Weaver, A. (2004) The Science of Climate Change. In H. Coward and A. Weaver (eds.) Hard Choices: Climate Change in Canada (Waterloo, ON: Wilfrid Laurier University Press).

Weiss, T. and M. Bramley (2009) Obama to Invest over Six Times More Per Capita in Renewable Energy and Energy Efficiency than Canada (Ottawa, ON: Pembina Institute).

Weiss, Th. (2009) What Happened to the Idea of World Government? International Studies Quarterly, Vol. 53, No. 2, pp. 253–271 (London: Wiley-Blackwell).

Weiss, Th. (2000) Governance, Good Governance and Global Governance: Conceptual and Actual Challenges, Third World Quarterly, Vol. 21, No. 5, pp. 795–814 (London: Routledge).

Weitzman, M. (1974) Prices vs. Quantities, Review of Economic Studies, Vol. 61, No. 4 (October), pp. 477–491 (London: Wiley-Blackwell).

Wennberg, J. (2010) State and Regional Climate Initiatives: Lessons for Federal Action (Washington, DC: Center for Climate Strategies)

Western Climate Initiative (2012a) North America 2050: A Partnership for Progress (Sacramento, CA: WCI) at http://www.westernclimateinitiative.org/document-archives/general/North-America-2050-Fact-Sheet/ (cited: 07/03/13).

Western Climate Initiative (2012b) Western Climate Initiative Jurisdictions Establish Non-profit Corporation to Support Greenhouse Gas Emissions Trading Programs (Sacramento, CA: WCI) http://www.westernclimateinitiative.org/news-and-updates (cited: 07/02/13).

Western Climate Initiative (2012c) Western Climate Initiative Status Update (Sacramento, CA: WCI) at http://www.westernclimateinitiative.org/news-and-updates (cited: 07/02/13).

Western Climate Initiative (2011) Western Climate Initiative Status Update, April 2011 (Sacramento, CA: WCI) at http://www.westernclimateinitiative.org/news-and-updates (cited: 12/05/12).

Western Climate Initiative (2010a) Design for the WCI Regional Program (Sacramento, CA: WCI) at http://westernclimateinitiative.org/component/remository/func-startdown/282/ (cited: 09/07/10).

Western Climate Initiative (2010b) Ensuring Offset Quality: Design and Implementation Criteria for a High-quality Offset Program

(Sacramento, CA: WCI) at http://www.rggi.org/docs/3_Regions_Offsets_Announcement_05_17_10.pdf (cited: 09/07/10).

Western Climate Initiative (2010c) Essential Requirements of Mandatory Reporting and the EPA Greenhouse Gas Reporting Program (Sacramento, CA: WCI) at http://www.westernclimateinitiative.org/news-and-updates/125-harmonization-of-essential-requirements-for-mandatory-reporting-in-us-jurisdictions-with-epa-mandatory-reporting-rule (cited: 09/07/10).

Western Climate Initiative (2010d) Harmonization of Essential Requirements of Mandatory Reporting in Canadian Jurisdictions with the WCI (Sacramento, CA: WCI) at http://www.westernclimateinitiative.org/news-and-updates/124-second-harmonization-package-for-reporting-requirements-for-canadian-jurisdictions-posted-for-stakeholder-comment (cited: 09/07/10).

Western Climate Initiative (2010e) Market Oversight Draft Recommendations (Sacramento, CA: WCI) at http://www.westernclimateinitiative.org/components/com_publiccomments/documents/Market_Oversight_Draft_Recommendations.pdf (cited: 09/07/10).

Western Climate Initiative (2010f) Program Design Presentation (Sacramento, CA: WCI) at http://www.westernclimateinitiative.org/component/remository/func-startdown/279/ (cited: 12/07/10).

Western Climate Initiative (2010g) The WCI Cap & Trade Program (Sacramento, CA: WCI) at http://www.westernclimateinitiative.org/the-wci-cap-and-trade-program (cited: 12/07/10).

Western Climate Initiative (2009a) The Climate Registry's Accreditation Program, WCI (Sacramento, CA: WCI) at http://www.westernclimateinitiative.org/component/remository/func-startdown/115/ (cited: 12/07/10).

Western Climate Initiative (2009b) WCI Offset Committee White Paper. Task 1: Offset System Essential Elements Offset Definition and Eligibility Criteria (Sacramento, CA: WCI) at http://www.westernclimateinitiative.org/component/remository/func-startdown/124/ (cited: 12/07/10).

Western Climate Initiative (2007) Statement of Regional Goal (Sacramento, CA: WCI) at http://www.westernclimateinitiative.org/component/remository/func-startdown/91/ (cited: 12/07/10).

Western Electricity Coordinating Council (2011) WECC Members (Salt Lake City, UT: Western Electricity Coordinating Council) at http://www.wecc.biz/About/Company/Pages/WECCMembers.aspx (cited: 07/18/11).

Weyland, Kurt (2001) Clarifying a Contested Concept: Populism in the Study of Latin American Politics, Comparative Politics, Vol. 34, No. 1, pp. 1–22 (New York: City University of New York) at http://www.jstor.org/stable/422412 (cited: 04/12/11).

Wikicommons (2010) Map: The Three Major and Two Minor NERC Interconnections, and the Nine NERC Regional Reliability Councils (Wikicommons.com) at http://en.wikipedia.org/wiki/File:Nercmap.JPG (cited: 10/28/10).

Winfield, M. (2002) Environmental Policy and Federalism. In H. Bakvis and G. Skogstad (eds.) Canadian Federalism: Performance, Effectiveness, and Legitimacy (Don Mills, ON: Oxford University Press).

Winfield, M., C. Demerse and J. Whitmore (2008) Climate Change and Canadian Energy Policy. In S. Bernstein, J. Brunnée, D. Duff and A. Green (eds.) A Globally Integrated Climate Policy for Canada (Toronto, ON: University of Toronto Press).

World Bank (2011) Emisiones de CO (toneladas métricas per cápita) (Washington, DC: Banco Mundial) at http://datos.bancomundial.org/indicador/EN.ATM.CO2E.PC+banco+mundial+emisiones+de+co2&cd=1&hl=es&ct=clnk&gl=mx (cited: 01/12/12).

World Business Council for Sustainable Development (2007) Establishing a Global Carbon Market: A Discussion on Linking Various Approaches to Create a Global Market (Geneva: World Business Council for Sustainable Development).

World Climate Report (2011) Biggest Drop in U.S. Greenhouse Gas Emissions (Washington, DC: World Climate Report) at http://www.worldclimatereport.com/index.php/2011/04/14/biggest-drop-in-us-greenhouse-gas-emissions/ (cited: 04/20/11).

World Mayors Summit on Climate (2010) Pacto de la Ciudad de México (Mexico City: World Mayors Summit on Climate) at http://www.wmsc2010.org/es/pacto-de-la-ciudad-de-mexico (cited: 03/12/10).

World Resources Institute (2010) U.S. State Climate Initiatives: MGGRA, RGGI & WCI (Washington, DC: World Resources Institute) at http://www.wri.org/chart/u-s-state-climate-initiatives-mggra-rggi-wci (cited: 09/07/12).

World Resources Institute (2008) Aggregate Contributions of Major GHG Countries (Washington, DC: World Resources Institute) at http://www.wri.org/chart/aggregate-contributions-major-ghg-emitting-countries (cited: 05/23/11).

Young, O. (1989) International Cooperation: Building Regimes for Natural Resources and the Environment (New York: Cornell University Press).

Young, O. (1992) The Effectiveness of International Institutions: Hard Cases and Critical Variables. In R. Rosenau and E. Czempiel (eds.) Governance without Government: Order and Change in World Politics (Cambridge: Cambridge University Press).

Zachary, P. and Ch. L. Carpentier (2003) Background Paper: Market-based Mechanisms for Carbon Sequestration, Energy Efficiency and Renewable Energy in North America—What Are the Options? (Montreal, QC: Commission for Environmental Cooperation).

Zanotti, L. (2005) Governmentalizing the Post-Cold War International Regime: The UN Debate on Democratization and Good Governance, Alternatives: Global, Local, Political, Vol. 30, No. 4 (October–December), pp. 461–487 (Ipswich, MA: EBSCO Publishing).

Zürn, M. (2010) Global Governance as Multi-level Governance. In H. Enderlein, S. Wälti and M. Zürn (eds.) Handbook on Multi-level Governance (Cheltenham, UK: Edward Elgar Publishing).

Index

Note: page numbers in italic type refer to Figures; those in bold type refer to Tables. Page numbers followed by 'n' and another number refer to footnotes.

THE INTERNATIONAL POLITICAL ECONOMY OF NEW REGIONALISMS SERIES

Other titles in the series

Comparative Regionalisms for Development in the 21st Century
Insights from the Global South
Edited by Emmanuel Fanta, Timothy M. Shaw and Vanessa T. Tang

Mapping Agency
Comparing Regionalisms in Africa
Edited by Ulrike Lorenz-Carl and Martin Rempe

The EU and the Eurozone Crisis
Policy Challenges and Strategic Choices
Edited by Finn Laursen

China-Africa Relations in an Era of Great Transformations
Edited by Li Xing with Abdulkadir Osman Farah

The European Union Neighbourhood
Challenges and Opportunities
Edited by Teresa Cierco

The New Democracy Wars
The Politics of North American Democracy Promotion in the Americas
Neil A. Burron

The European Union after Lisbon
Polity, Politics, Policy
Edited by Søren Dosenrode

Roads to Regionalism
Genesis, Design, and Effects of Regional Organizations
Edited by Tanja A. Börzel, Lukas Goltermann, Mathis Lohaus and Kai Striebinger

New Regionalism or No Regionalism?
Emerging Regionalism in the Black Sea Area
Edited by Ruxandra Ivan

Our North America
Social and Political Issues beyond NAFTA
Edited by Julián Castro-Rea

Community of Insecurity
SADC's Struggle for Peace and Security in Southern Africa
Laurie Nathan

Global and Regional Problems
Towards an Interdisciplinary Study
Edited by Pami Aalto, Vilho Harle and Sami Moisio

The Ashgate Research Companion to Regionalisms
Edited by Timothy M. Shaw, J. Andrew Grant and Scarlett Cornelissen

Asymmetric Trade Negotiations
Sanoussi Bilal, Philippe De Lombaerde and Diana Tussie

The Rise of the Networking Region
The Challenges of Regional Collaboration
in a Globalized World
Edited by Harald Baldersheim,
Are Vegard Haug and Morten Øgård

Shifting Geo-Economic Power of the
Gulf Oil, Finance and Institutions
Edited by Matteo Legrenzi
and Bessma Momani

Building Regions
The Regionalization of the World Order
Luk Van Langenhove

National Solutions to
Trans-Border Problems?
The Governance of Security and Risk
in a Post-NAFTA North America
Edited by Isidro Morales

The Euro in the 21st Century
Economic Crisis and Financial Uproar
María Lorca-Susino

Crafting an African Security Architecture
Addressing Regional Peace and Conflict
in the 21st Century
Edited by Hany Besada

Comparative Regional Integration
Europe and Beyond
Edited by Finn Laursen

The Rise of China
and the Capitalist World Order
Edited by Li Xing

The EU and World Regionalism
The Makability of Regions
in the 21st Century
Edited by Philippe De Lombaerde
and Michael Schulz

The Role of the European Union in Asia
China and India as Strategic Partners
Edited by Bart Gaens, Juha Jokela
and Eija Limnell

China and the Global Politics
of Regionalization
Edited by Emilian Kavalski

Clash or Cooperation of Civilizations?
Overlapping Integration and Identities
Edited by Wolfgang Zank

New Perspectives on Globalization
and Antiglobalization: Prospects for a
New World Order?
Edited by Henry Veltmeyer

Governing Regional Integration for
Development: Monitoring Experiences,
Methods and Prospects
Edited by Philippe De Lombaerde,
Antoni Estevadeordal and Kati Suominen

Europe-Asia Interregional Relations
A Decade of ASEM
Edited by Bart Gaens

Cruising in the Global Economy
Profits, Pleasure and Work at Sea
Christine B.N. Chin

Beyond Regionalism?
Regional Cooperation, Regionalism and
Regionalization in the Middle East
Edited by Cilja Harders
and Matteo Legrenzi

The EU-Russian Energy Dialogue
Europe's Future Energy Security
Edited by Pami Aalto

Regionalism, Globalisation
and International Order
Europe and Southeast Asia
Jens-Uwe Wunderlich

EU Development Policy
and Poverty Reduction
Enhancing Effectiveness
Edited by Wil Hout

An East Asian Model for Latin
American Success
The New Path
Anil Hira

European Union and New Regionalism
Regional Actors and Global Governance
in a Post-Hegemonic Era
Second Edition
Edited by Mario Telò

Regional Integration and Poverty
*Edited by Dirk Willem te Velde
and the Overseas Development Institute*

Redefining the Pacific?
Regionalism Past, Present and Future
*Edited by Jenny Bryant-Tokalau
and Ian Frazer*

Latin America's Quest for Globalization
The Role of Spanish Firms
*Edited by Félix E. Martín
and Pablo Toral*

Exchange Rate Crises
in Developing Countries
The Political Role of the Banking Sector
Michael G. Hall

Globalization and Antiglobalization
Dynamics of Change in the
New World Order
Edited by Henry Veltmeyer

Twisting Arms and Flexing Muscles
Humanitarian Intervention and
Peacebuilding in Perspective
*Edited by Natalie Mychajlyszyn
and Timothy M. Shaw*

Asia Pacific and Human Rights
A Global Political Economy Perspective
Paul Close and David Askew

Demilitarisation and Peace-Building
in Southern Africa
Volume I – Concepts and Processes
*Edited by Peter Batchelor
and Kees Kingma*

Demilitarisation and Peace-Building
in Southern Africa
Volume II – National and
Regional Experiences
*Edited by Peter Batchelor
and Kees Kingma*

Persistent Permeability?
Regionalism, Localism, and Globalization
in the Middle East
*Edited by Bassel F. Salloukh
and Rex Brynen*

The New Political Economy of
United States-Caribbean Relations
The Apparel Industry and the
Politics of NAFTA Parity
Tony Heron

The Nordic Regions and the European Union
*Edited by Søren Dosenrode
and Henrik Halkier*

For Product Safety Concerns and Information please contact our EU
representative GPSR@taylorandfrancis.com
Taylor & Francis Verlag GmbH, Kaufingerstraße 24, 80331 München, Germany

www.ingramcontent.com/pod-product-compliance
Ingram Content Group UK Ltd.
Pitfield, Milton Keynes, MK11 3LW, UK
UKHW010959180926
475701UK00004B/450